DATE DUE		
NOV 04 '91 S		
NOV 1 5 1993 S		
DEC 1 9 1997 S		
JAN 1 4 1998		
WITHDRAWN		

ATOM

ATOM Journey Across the Subatomic Cosmos

Illustrated by D. F. Bach

ISAAC ASIMOV

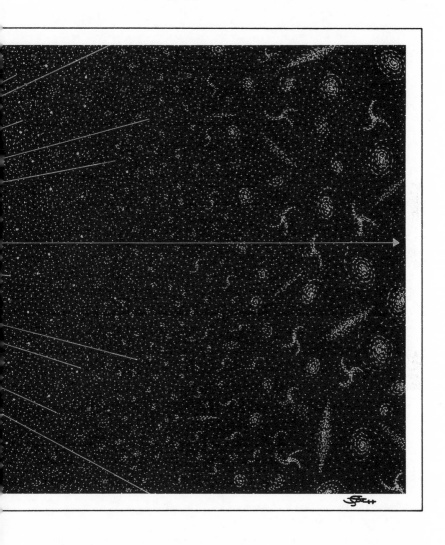

T·T TRUMAN TALLEY BOOKS / **DUTTON** / NEW YORK

DUTTON
Published by the Penguin Group
Penguin Books USA Inc., 375 Hudson Street,
New York, New York 10014, U.S.A.
Penguin Books Ltd, 27 Wrights Lane,
London W8 5TZ, England
Penguin Books Australia Ltd, Ringwood,
Victoria, Australia
Penguin Books Canada Ltd, 2801 John Street,
Markham, Ontario, Canada L3R 1B4
Penguin Books (N.Z.) Ltd, 182–190 Wairau Road,
Auckland 10, New Zealand

Penguin Books Ltd, Registered Offices:
Harmondsworth, Middlesex, England

First published by Truman Talley Books · Dutton,
an imprint of New American Library,
a division of Penguin Books USA Inc.
Distributed in Canada by McClelland & Stewart Inc.

First Printing, May, 1991
10 9 8 7 6 5 4 3 2 1

LIBRARY OF CONGRESS CATALOGING-IN-PUBLICATION DATA
Asimov, Isaac, 1920–
    Atom : journey across the subatomic cosmos / Isaac Asimov.
        p. cm.
    "Truman Talley books."
    ISBN 0-525-24990-7
    1. Atoms. I. Title.
QC173.A778   1991
539.7—dc20   90-21343
                                                        CIP
Printed in the United States of America
Set in Century Expanded

To Truman "Mac" Talley
Who represents book publishing at its best.

CONTENTS

1

MATTER

Dividing Matter

Suppose you had a large heap of small, smooth pebbles—thousands of them. If you had nothing better to do, you might decide to divide it into two smaller heaps, approximately equal in size. You could discard one of these heaps, keep the other, and divide it in two again. Of these two still smaller heaps, you could discard one and keep the other for further division, and repeat the process over and over.

You might wonder how long you could keep that up. Forever? You know better than that. No matter how large the heap was to begin with, you would eventually be left with a tiny "heap" made up of just two pebbles. (This would happen surprisingly quickly. Even if you started with a million pebbles, you would be down to two pebbles after about twenty divisions.) If you divided a heap of two pebbles once again, you would end up with one heap consisting of a single pebble, and the game would be over. You can't divide one pebble.

But wait! You can. You could place the pebble on an anvil and pound it with a hammer. You would break it up into fragments, and you could divide this into smaller and smaller heaps until you were down to a single fragment. You could then pound the fragment into dust and then divide the heap of dust until you ended up with a single, hardly visible dust particle. You could break that up, and keep on going.

This is not really a practical game because it's very hard to handle a grain of dust and try to break it up further. But you can *imagine*. Imagine that you can break up the dust into still finer particles, which you can break up yet further, getting it ever finer. Now ask yourself: is there any end to this?

It might not seem like a very important question, or even a particularly sensible question, in that you can't really try the experiment in any practical way. You find yourself dealing, very quickly, with objects that are too small to see, so that you don't even know whether or not you're breaking the heap down any further. Nevertheless, certain ancient Greek philosophers asked themselves this question and started a chain of thought that is still occupying thinkers to this day, twenty-five centuries later.

The Greek philosopher Leucippus (490– ? B.C.) is the first person we know of by name supposed to have considered this problem of dividing matter, and to have come to the conclusion that the process could *not* continue forever. He insisted that, sooner or later, one had to reach a fragment of matter so small that it could not be broken down into anything smaller.

A younger man, Democritus (460–370 B.C.), was one

2

of Leucippus's pupils. He accepted the notion of fragments of matter so small as to be unbreakable. He called such fragments *atomos*, which in Greek means "unbreakable," and such a fragment has come to be called an atom in English. To Democritus, all matter consisted of a collection of atoms, and if there was space between the atoms, that space contained nothing (the "void").

Democritus is supposed to have written sixty books expounding his theories, including his notions of what we now call atomism. In those days, though, when there was no printing and all books had to be hand copied, there were hardly ever very many copies; and, partly because his views were unpopular, the books were not copied many times. Over the centuries many books vanished. None of Democritus's books has survived.

Most philosophers of the time felt that it didn't make sense to suppose that some tiny individual particle was indivisible. They thought it made more sense to suppose that everything could be broken up into smaller and smaller bits of matter, endlessly.

In particular, the Greek philosophers Plato (ca. 427–347 B.C.) and Aristotle (384–322 B.C.) didn't accept atoms. Because they were the most profound and mentally wide-ranging of the ancient philosophers, their views tended to carry the day. But the argument was not unanimous. The influential Greek philosopher Epicurus (341–270 B.C.) took up atomism as the central core of his teachings. Epicurus is supposed to have written 300 books (ancient books tended to be short, incidentally), but none of them has survived.

The most important of the Epicureans in this connection was a Roman, Titus Lucretius Carus (96–55 B.C.), usually known simply as Lucretius. In 56 B.C., he published a long poem in Latin entitled *De Rerum Natura* (Latin for *On the Nature of Things*). In it, he explained the Epicurean view of atomism in great detail.

The book was very popular in its time, but in later

centuries, after Christianity had grown popular, Lucretius was denounced for what was considered to be atheism. He was no longer copied, and what copies already existed were destroyed or lost. Even so, one copy (only one!) survived through the Middle Ages and was discovered in 1417. It was recopied and then, half a century later, when printing came into use, Lucretius's poem was one of the first items to be printed.

The poem spread throughout western Europe and was the chief source of knowledge of the ancient theories of atomism. The French philosopher Pierre Gassendi (1592–1655), having read Lucretius, adopted the atomistic view himself, and wrote it up persuasively, thus spreading the doctrine.

In all the two thousand years between Leucippus and Gassendi, however, atomism, pro and con, was simply a subject of endless discussion among scholars. There was no *evidence* either for or against atomism. Various scholars accepted atoms or rejected them, according to which point of view pleased them better, or seemed more sensible. There was no way of forcing one view on someone who held the other view firmly. It was a subjective decision, and there was no arguing with taste.

About this time, however, some scholars were beginning to perform experiments; to set questions to nature, so to speak, and to study the results. In this way, evidence could be produced that was scientifically "compelling"; that is, it was evidence that compelled others to accept a point of view that they were subjectively against (provided they were intellectually honest).

The first to perform experiments that seemed to have a connection with the question of atomism was the British scientist Robert Boyle (1627–1691), who was strongly influenced by Gassendi's writings, and who was consequently an atomist.

In 1662, Boyle made use of a glass tube shaped like the letter J. The short arm was closed and the long arm open. He poured mercury into the opening and it filled the bottom, trapping air in the closed short arm. Boyle then poured additional mercury into the tube, the weight of which compressed the air in the short arm, decreasing the volume of the air as a result. If he doubled the height of the mercury column in the long arm, the volume of air in the short arm was halved. When the mercury was removed and the pressure released, the volume of air increased. This inverse relationship between pressure and volume has been called Boyle's law ever since.

This behavior of air under pressure was easily explained if one made use of atoms. Suppose the air is made up of atoms that are widely separated, with nothing in between—as Democritus had suggested. (This would account for the fact that a volume of air weighs so much less than the same volume of water or marble, where the atoms might be in contact.) Placing the air under pressure would force the atoms close together, squeezing out some of the nothingness, so to speak, and would decrease the volume. Relieving the pressure would allow the atoms to spread outward.

For the first time, atomism began to gain an upper hand. Someone might think that it wasn't sensible, or perhaps that it wasn't esthetic, to suppose the existence of atoms, but one could not argue with Boyle's experiment. This was especially true in that anyone could run the experiment himself and come up with the same observations.

If we must accept Boyle's experiment, then atomism offers a simple and logical explanation of his findings. Explaining the results without atoms is much more difficult.

From that point on, then, more and more scientists were atomists, but the issue was not yet completely settled. (We'll get back to the subject later.)

Elements

The ancient Greek philosophers wondered what the world was made of. Clearly, it was made of innumerable types of things, but scientists have always felt the urge to simplify. There was the feeling, therefore, that the world was made of some basic material (or some very few basic materials), of which everything else was one variation or another.

Thales (ca. 640–546 B.C.) is the first Greek philosopher supposed to have suggested that water was the basic material out of which everything was formed. Another, Anaximenes (570–500 B.C.), thought it was air. Still another, Heraclitus (ca. 535–475 B.C.), thought it was fire, and so on.

There was no way of deciding among these suggestions for there was no real evidence one way or another. The Greek philosopher Empedocles (495–435 B.C.) settled the issue by compromise. He suggested that the world was made of several different basic substances: fire, air, water, and earth. To this Aristotle added aether (from a Greek word for "blazing") as a special substance out of which the luminous heavenly bodies were composed.

These basic substances are called elements in English, from a Latin word of unknown origin. (We still describe storms by speaking of "the raging of the elements" as water pours down, air blows about, and fire burns as lightning.)

To those people who accepted the notion of the various elements, and who were atomists, it made sense to suppose that each element was composed of a different type of atom, so that the world consisted of four different types of atoms altogether, with a fifth type for the heavenly aether.

Even with only four types of atoms, it was possible to account for the great variety of objects on Earth. One only had to imagine that the various substances were made up

6

of combinations of different numbers of different types of atoms in different arrangements. After all, with only twenty-six letters (or with just two symbols, a dot and a dash), it is possible to build up hundreds of thousands of different words in English alone.

However, the doctrine of the four (or five) elements began to fade even as atomism began to move ahead. In 1661, Boyle wrote a book, *The Skeptical Chemist*, in which he took up the position that it was useless to guess at what the basic substances of the world might be. One had to determine what they were by experiment. Any substance that could not be broken down by chemical manipulation into any simpler substance was an element. Any substance that *could* be broken down into simpler components was not an element.

This is indisputable in principle, but it is not entirely easy in practice. Some substances cannot be broken down into anything simpler and might seem to be elements, but then the time might come when advances in chemistry will make it possible to break them down. And again, when one substance is converted into another it isn't always easy to decide which of the two is simpler.

Nevertheless, beginning with Boyle and continuing for over three centuries, chemists have labored to find substances that can be identified as elements. Examples of familiar substances that have been recognized as elements in this way are gold, silver, copper, iron, tin, aluminum, chromium, lead, and mercury. Gases such as hydrogen, nitrogen, and oxygen are elements. Air, water, earth, and fire are *not* elements.

At the present time, 106 elements are known. Eighty-three of them occur naturally on Earth in reasonable quantities, and the remaining twenty-three occur either in traces or only after having been manufactured in a laboratory. This means there are 106 different types of atoms known.

Atomism Triumphant

Most substances as they occur on Earth are not elements, but can be broken down into the various elements that make them up. Those substances that are put together out of a combination of elements are known as compounds (from Latin words meaning "to put together").

Chemists grew increasingly interested in trying to determine how much of each element might exist in a particular compound. Beginning in 1794, the French chemist Joseph Louis Proust (1754–1826) worked on this problem, and made a crucial discovery. There is a compound we now call copper carbonate. Proust began with a pure sample of this substance and broke it down into the three elements that made it up: copper, carbon, and oxygen. He found, in 1799, that in every sample he worked with, no matter how it was prepared, there were present for every five parts of copper (by weight) four parts of oxygen and one part of carbon. If he added additional copper to the mixture in preparing copper carbonate, the additional copper was left over. If he began with a shortage of copper, only the proportionate amount of carbon and oxygen combined with it to form copper carbonate, and the rest of the carbon and oxygen was left over.

Proust showed that this was also true for a number of other compounds he worked with. The elements of which they were composed were always present in definite proportions. This was called the law of definite proportions.

The law of definite proportions offered strong support for atomism. Suppose, for instance, that copper carbonate is made up of little groups of atoms (called molecules, from Latin words meaning "a small mass"), each group consisting of one copper atom, one carbon atom, and three oxygen atoms. Suppose also that the three oxygen atoms, taken

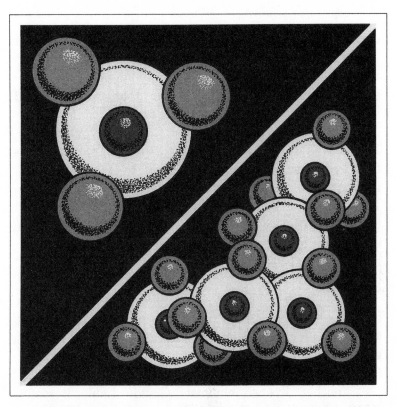

In 1799, Joseph Louis Proust broke down a pure sample of a sub-stance into the three elements that comprised it: copper, carbon, and oxygen. He found that in every sample he worked with there were present for every five parts of copper (by weight) four parts of oxygen and one part of copper. The atomic elements that compose a compound are always present in definite proportions.

together, are four times as heavy as the carbon atom, and that the copper atom is five times as heavy as the carbon atom. If every molecule of that compound is made up of that combination, then copper carbonate would always be made up of five parts copper, four parts oxygen, and one part carbon.

If it were possible to include in the molecule 1½ atoms of copper, or 3⅓ atoms of oxygen, or only ⅚ of an atom of

9

carbon, the proportions of the three substances might vary from sample to sample of copper carbonate. However, the proportions *don't* vary. This not only supports the idea of atoms, but Democritus's suggestion that an atom is indivisible. It exists as an intact piece or as nothing.

The difference between the work of Democritus and Proust was this, however: Democritus had only a suggestion; Proust had *evidence*. (This is not to be taken as meaning that Proust was necessarily a greater or wiser man than Democritus. Proust had the benefit of twenty-one additional centuries of thought and work that he could draw upon. You might easily argue that it was much more remarkable that Democritus could hit on the truth so early in the game.)

Even with evidence, Proust did not necessarily have it all his own way. After all, it was possible that Proust's analyses were wrong, or that he was so eager to prove his own idea that he unconsciously twisted his observations. (Scientists are only human, and such things happen.)

Another French chemist, Claude Louis Berthollet (1748–1822), fought Proust every step of the way. He insisted that *his* analyses showed that compounds could be made up of elements in varying proportions. In 1804, however, the Swedish chemist Jöns Jakob Berzelius (1779–1848) began meticulous analyses that backed Proust's notion, and proved to the chemical world that the law of definite proportions was right.

At the same time, the English chemist John Dalton (1766–1844) was also working on the problem. He found that it was possible for compounds to be made up of elements in widely different proportions. Thus, in one gas, with molecules made up of carbon and oxygen, the proportions were three parts carbon to four parts oxygen. In another gas, with molecules made up of carbon and oxygen, the proportions were three parts carbon to eight parts ox-

10

ygen. These, however, were two different gases with two different sets of properties, and for each one the law of definite proportions held.

Dalton suggested that in one gas the molecule was made up of an atom of carbon and an atom of oxygen, whereas in the other it was made up of an atom of carbon and *two* atoms of oxygen. It eventually turned out that he was correct, and the two gases came to be called carbon monoxide and carbon dioxide, respectively. (The prefix mon- is from the Greek word for "one," and di- is from the Greek word for "two.")

Dalton found this sort of thing was true in other cases, and in 1803 he announced this as the law of multiple proportions. He pointed out that this fit the notion of atoms, and it was he who called them atoms, deliberately going back to the old term as a tribute to Democritus.

Dalton said that to account for what was being found out about the proportion of elements contained in compounds, one had to decide that each element is made up of a number of atoms, all with the same fixed mass; that different elements have atoms of different masses; and that molecules are made up of a small, fixed number of different intact atoms.

In 1808, Dalton published a book entitled *New System of Chemical Philosophy*, in which he gathered all of the evidence he could find in favor of atomism and showed how it all fit together. With this book, Dalton established the modern atomic theory—modern, as opposed to that of the Greeks.

As it happens, the word *theory* is not properly understood by the general public, which tends to think of a theory as a "guess." Even dictionaries do not properly describe what the word means to scientists.

Properly speaking, a theory is a set of basic rules, supported by a great many confirmed observations by many scientists, that explains and makes sensible a large number of facts that, without the theory, would seem to be unconnected. It is as though the facts and observations are a number of dots representing cities, and lines representing country and state boundaries, distributed higgledy-piggledy on paper, making no sense. A theory is a *map* that puts each dot and line into the right place and makes a connected and sensible picture out of it all.

Theories are not necessarily correct in every detail, to begin with, and might never be entirely correct in every detail, but they are sufficiently correct (if they are good theories) to guide scientists in understanding the subject the theory deals with, in exploring further observations, and, eventually, in improving the theory.

Each of the basic rules Dalton set up for his atomic theory was not *quite* right. It turned out, eventually, that an element could have atoms of different mass, that two elements might have some atoms that were of the same mass, and that not all molecules were made up of small numbers of atoms. Dalton's rules were sufficiently close to right, however, to be very useful, and, as chemists learned more and more about atoms, they were able to correct the rules, as we shall see later on.

No scientific theory is instantly accepted by scientists. There are always those scientists who are suspicious of anything new—and this is perhaps a good thing. Theories should not slide into acceptance too easily; they should be questioned and tested vigorously. In this way, weak spots in the theory will be uncovered and, perhaps, strengthened.

As it happened, some of the most eminent chemists of Dalton's day were suspicious of the new theory, but it turned out to be so useful in helping to understand the observations of chemistry that chemist after chemist fell into line, and the entire scientific world eventually became atomists.

The Reality of Atoms

However well atomic theory worked, and however ingeniously it was improved, and however it managed to point the way to new discoveries, one disturbing fact remained: no one could see atoms or detect them in any way. All of the evidence in favor of atoms was indirect. You *inferred* that they existed from this fact, and *deduced* that they existed from that observation, but all of the inferences and deductions might be wrong. Atomic theory seemed to set up a scheme that worked, but it might have been just a simple model for something that was actually much more complicated. The working mode of the time was analogous to playing poker with chips. The chips can be used to bet with and to show how much money is being lost and won, and will be absolutely accurate in every way—but those chips are *not* money. They just symbolize the money.

Suppose, then, that the idea of atoms is merely a case of playing chemistry with chips. Atomism worked, but the atoms merely represented a truth that was much more complicated. There were some chemists, even a hundred years after Dalton, who were cautiously aware of this, and who warned against taking atoms too literally. Use them by all means, they would say, but don't think that they are necessarily really there in the shape of minute billiard balls. One scientist who thought this way was the Russian-German chemist Friedrich Wilhelm Ostwald (1853–1932).

The answer to this problem had long been on the way, however, and it started with an observation that seemed to have nothing to do with atoms, by a scientist who wasn't interested in atoms. (It's important to remember that all knowledge is of a piece and that any observation can have an unexpected and surprising connection to something that apparently has nothing to do with it.)

13

The vibration of a grain of pollen in water demonstrates the movement of the invisible molecules of water surrounding it.

In 1827, the Scottish botanist Robert Brown (1773–1858) was using a microscope to study pollen grains suspended in water. He noticed that each pollen grain was moving slightly and erratically, first in one direction then in another, as though it were shivering. He made sure that this wasn't the result of currents in the water or of motions set up by the fact that the water was evaporating. Brown concluded it had to be something else that caused the movement.

Brown tried other types of pollen, found that all of the grains moved in this fashion, and wondered if it was because the pollen grains had the spark of life in them. He tried

14

pollen grains from herbariums, grains that were at least a century old. They moved in just the same way. He went on to try small objects in which there was no question of life existing—bits of glass, coal, or metals—and they *all* moved. This came to be called Brownian motion, and no one, at first, could explain it.

In the 1860s, however, the Scottish mathematician James Clerk Maxwell (1831–1879) tried to explain the behavior of gases on the basis that the atoms and molecules that made them up were in constant motion. Such constant motion of atoms had been suspected by early atomists, but Maxwell was the first to succeed in working the theory out mathematically. The way in which moving atoms and molecules bounced off each other, and off the walls of a container, as mathematically modeled by Maxwell, completely explained the behavior of gases. It explained Boyle's law, for instance.

Maxwell's work also produced a new understanding of temperature, for it turned out that temperature was the measure of the average speed of motion of the atoms and molecules making up not only gases, but liquids and solids. Even in solids, where atoms or molecules are frozen in place and can't move bodily from one point to another, those atoms or molecules vibrate about their average position, and the average speed of vibration represents the temperature.

In 1902, the Swedish chemist Theodor Svedberg (1884–1971) pointed out that one might explain Brownian motion by supposing that an object in water is bombarded from all sides by moving water molecules. Ordinarily, the bombardment from all sides is equal, so that the object remains at rest. To be sure, by sheer chance, a few more molecules might strike from one direction or another, but so many molecules strike all together that a small deviation from exact equality (two or three out of trillions) does not produce measurable movement.

If an object suspended in water is very small, however,

the number of molecules striking it from all sides is comparatively small, too, and if there is a small deviation, that might represent a fairly large effect, comparatively. The particle responds to the push of a few extra molecules from one particular direction by jerking slightly in the direction of the push. In the next moment, there are extra collisions in another direction, and the particle is pushed in that new direction. The particle moves randomly and erratically in response to the random motion of the surrounding molecules.

Svedberg was only speculating, but in 1905, the German-Swiss mathematician Albert Einstein (1879–1955) applied Maxwell's theory to the bombardment of small particles and showed quite conclusively that those particles would jiggle exactly as the pollen grains were observed to do. In other words, he presented mathematical equations that described Brownian motion.

In 1908, the French physicist Jean Baptiste Perrin (1870–1942) set about checking Einstein's equations against actual observations. He placed a fine powder of gum resin in water. If there were no bombardment by water molecules, then all of the particles of gum resin ought to have gone to the bottom of the container and remained there. If there were bombardment, some of the particles would be kicked upward against the pull of gravity. To be sure, those particles would settle again, but they would then be kicked up again, too. Some that were already up would be kicked up still further.

At any given time, the particles of gum resin would be spread upward. Most would be at the bottom, but some would be a little distance above, a few a greater distance above, still fewer a still greater distance above, and so on.

The mathematical equation worked out by Einstein showed what numbers of particles there should be at every height, depending upon the size of the particles and the size of the water molecules striking them. Perrin counted the number of particles at various heights and found that they followed Einstein's equation exactly. From this he calculated what size the water molecules must be, and what size the atoms that made them up must be.

Perrin published his results in 1913. The atoms, he had calculated, were roughly a hundred-millionth of a centimeter across. Put it another way: 100 million atoms placed side by side would stretch across a centimeter (250 million atoms placed side by side would stretch across an inch).

This was the nearest thing yet to an actual observation of atoms. If they could not quite be seen, the effects of their collisions could be seen and their actual size could finally be worked out. The most hard-nosed scientists had to give in. Even Ostwald admitted that atoms were real, that they weren't just make-believe models.

In 1936, the German physicist Erwin Wilhelm Mueller (1911–1977) got the idea of a device that would make it possible to magnify the point of a fine needle to such an extent that one could make pictures of it, with the atoms that compose it lined up as little luminous dots. By 1955, such atoms could actually be seen.

Yet people still speak of the atomic *theory*, because that is what it is—an intellectual map of large aspects of science that can be neatly explained by the existence of atoms. A theory, remember, is not a "guess," and no sane and qualified scientist can doubt that atoms exist. (This aspect of the proof that atoms exist is also true of other well-established scientific theories. The fact that they are theories does not make them uncertain, even when various fine details are still under dispute. This is particularly true of the theory of evolution, which is under constant attack from people who are either ignorant of science or, worse, who allow their superstitions to overcome what knowledge they might have.)

The Differences Among Atoms

It seems reasonable to suppose that if there are different types of atoms they must differ among themselves, somehow, in their properties. If this were not so, and if all atoms were identical in their properties, then why should some atoms, when heaped together, form gold, while others formed lead?

The ancient Greeks had their greatest intellectual success with the development of a rigorous form of geometry, so it was natural for some among them to think in terms of shapes when they thought of the atoms making up their "elements." To the Greeks, atoms of water might be viewed as spherical bodies that slipped over each other easily, which was why water poured. Atoms of earth would be cubic and stable so that earth *didn't* flow. Atoms of fire would be jagged and sharp, which was what made fire so painful, and so on.

The ancient Greeks also did not have it quite clear in their mind that one type of atom did not change into another. This was especially true if you considered that gold and lead were both varieties, in the main, of the element earth. Perhaps it was only necessary to pull apart the earth atoms in lead and put them into another arrangement that would make them gold; or one might modify the earth atoms in lead to change them slightly into a form that would make them gold.

For about two thousand years, various people, some of whom were earnest and science-minded, while a great many others were outright fakers and charlatans, kept trying to change base metals such as lead into the noble metal gold. This is called transmutation, from Latin words meaning "to change across." They always failed.

By the time the modern atomic theory was advanced,

18

To the ancient Greeks, atoms of water might be viewed as spherical bodies that slipped water over each other easily, which was why water poured.

it seemed clear that atoms were not only different from each other, but that one type of atom could not be changed into another. Each atom was fixed and permanent in its properties, so that an atom of lead could not be changed into an atom of gold. (The time was to come, as we shall see, when this was found to be not *quite* true, under very special conditions.)

But if different types of atoms are different from one another, of just what does the difference consist? Dalton reasoned as follows. If the water molecule is made up of eight parts oxygen to one part hydrogen, and if the molecule is made up of one atom of oxygen and one atom of hydrogen, then it must be that the individual oxygen atom weighs eight times as much as the individual hydrogen atom. (To be more precise, one should say that the individual oxygen atom has eight times the "mass" of the individual hydrogen atom. The weight of an object is the force with which the Earth attracts it, whereas the mass of an object is, roughly speaking, the amount of matter it contains. Mass is the more fundamental of the two concepts.)

Of course, Dalton had no way of knowing the mass of either a hydrogen or an oxygen atom, but whatever it was, the oxygen atom had a mass eight times that of a hydrogen atom. You could say that a hydrogen atom had a mass of 1, without saying 1 *what*. You could then say that an oxygen atom has a mass of 8. (Actually, we now say the hydrogen atom is 1 dalton, in honor of the scientist, but it is customary simply to leave it as 1.)

Dalton went to work with compounds containing other elements and worked out a system of numbers representing the relative masses of them all. He called them *atomic weights*, and the term is still used today, even though we should speak of atomic masses. (It frequently happens that scientists begin to use a particular term and then decide that another term would have been better, but find it is too late to change because people have grown far too accustomed to the poorer term. We'll come across other cases of the sort in this book.)

The trouble with Dalton's method of determining atomic weights was that he was forced to make assumptions that could too easily be wrong. He assumed that a water molecule consisted of one atom of hydrogen and one of oxygen, but he didn't have any evidence for that.

In that case, one must look for evidence. In 1800, the British chemist William Nicholson (1753–1815) passed an electric current through acidified water and obtained bubbles of both hydrogen and oxygen. Continued investigation of this phenomenon showed that the volume of hydrogen formed was just twice that of the oxygen, although the mass of oxygen liberated was eight times the mass of the double volume of hydrogen.

Why was twice the volume of hydrogen produced as compared to oxygen? Could it be that the water molecule was composed of *two* hydrogen atoms and one oxygen atom, instead of one of each? Could it be that the oxygen atom was eight times as heavy as both hydrogen atoms put together, or sixteen times as heavy as a single hydrogen atom? In other words, if hydrogen had an atomic weight of 1, was the atomic weight of oxygen 16, rather than 8?

Dalton refused to accept this notion. (It often happens that a great scientist, having taken a giant step forward, refuses to take other steps—as though the great first effort had exhausted him—and leaves it to others to continue to march forward.)

In this case, it was Berzelius who took the forward step, placing hydrogen at 1 and oxygen at 16. He continued with other elements and, in 1828, published a table of atomic weights that was much better than Dalton's had been. From the work of Berzelius, it seemed clear that every element had a different atomic weight, and that each atom of a particular element had the same atomic weight. (I must remind you again that these conclusions eventually proved to be not quite right, but they were near enough to right to be useful to chemists for nearly a century. Eventually, as more knowledge was gained, these views were modified in ways that slightly changed and immeasurably strengthened the atomic theory. This improvement of theories happens over and over and is the pride of science. To suppose that this should not happen and that theories should be absolutely correct to begin with is to suppose that a stairway stretching upward for five stories should consist of a single five-story-high step.)

Well, then, the volume of hydrogen produced when water is broken down by an electrical current is twice the volume of oxygen. How do we know from this that there are two hydrogen atoms to one oxygen atom in the molecule? It seemed sensible to Berzelius to suppose so, but he didn't know for sure. It, too, was an assumption, even though there was more evidence behind it than there was behind Dalton's assumption that there was one hydrogen atom and one oxygen in the water molecule.

In 1811, the Italian physicist Amedeo Avogadro (1776–1856) made a more general assumption. He suggested that in the case of *any* gas, a given volume always contains the same number of molecules. If one gas has twice the volume of another gas, the first gas has twice as many molecules as the other. This is called Avogadro's hypothesis. (A hypothesis is an assumption that is sometimes advanced just to see what the consequences would be. If the consequences go against known observations, then the hypothesis is wrong and it can be dismissed.)

Naturally, when a competent scientist advances a hypothesis that he thinks might be true, there is a good chance it will turn out to be true. One way of testing Avogadro's hypothesis, for instance, is to study a great many gases and to work out the number of each of the different types of atoms in the molecules of those gases on the basis that the hypothesis is true.

If one does that and ends by violating known observations, or ends by producing a contradiction—as when one line of argument based on the hypothesis shows that a particular molecule must have a certain atomic composition, and another line of argument shows it must have a different atomic composition—then Avogadro's hypothesis would have to be thrown out.

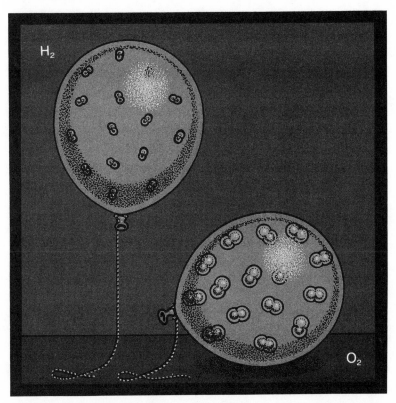

Avogadro's law: equal volumes of all gases under identical conditions of temperature and pressure contain equal numbers of molecules. For example, it might take .1 gram of hydrogen gas to fill a child's balloon. It would take approximately 1.6 grams of oxygen gas to inflate an identical balloon to an equal size, but both balloons would contain approximately the same number of molecules.

Actually, no one has ever found a case in which Avogadro's hypothesis is truly misleading, and the theory is no longer a hypothesis but is considered a fact, although there are conditions under which it must be modified. It is still *called* Avogadro's hypothesis, however, because chemists are so accustomed to calling it that.

One problem, however, was that when Avogadro's hypothesis was first advanced, very few chemists paid any attention to it. They either didn't hear of it, or dismissed it as either ridiculous or unimportant. Even Berzelius didn't make use of the hypothesis, so that his table of atomic weights was wrong in places.

In 1858, however, the Italian chemist Stanislao Cannizzaro (1826–1910) came across Avogadro's hypothesis and saw that that was what was needed to make sense out of figuring out how many atoms of each element there were in a compound, and getting the correct figures for atomic weight.

In 1860, there was a great international chemical congress, which chemists from all over Europe attended (it was the first of such international congresses). At that congress, Cannizzaro convincingly explained the hypothesis.

This at once improved the entire notion of atomic weight. About 1865, the Belgian chemist Jean-Servais Stas (1813–1891) put out a new table of atomic weights that was better than Berzelius's. Forty years later, the American chemist Theodore William Richards (1868–1928) made even more refined observations and got the very best values one could get before (as we shall see) the entire subject of atomic weight had to be modified because of new discoveries. By Richards's time, Nobel prizes were being handed out, and for his work on atomic weights he got the Nobel prize for chemistry in 1914.

As it happens, the element with the lowest atomic weight is hydrogen. If its atomic weight is set arbitrarily at 1, then the atomic weight of oxygen is a little bit less than 16. (That it is not exactly 16 is a point we will consider later on.) However, oxygen easily combines with a great many other elements, and it is much simpler to compare the atomic weight of some particular element to oxygen than to hydrogen. It is convenient, then, to set oxygen's atomic weight at some exact figure. It shouldn't be set at

1 because that would give seven elements atomic weights of less than 1, which would be inconvenient in making chemical calculations.

It became customary, then, to set the atomic weight of oxygen at exactly 16, which made the atomic weight of hydrogen just a little bit greater than 1. That meant that no element had an atomic weight of less than 1. Stas's list was made that way and that set the fashion. (However, the situation has been changed very slightly in recent years for reasons that will be explained later.)

If the elements are listed in order of increasing atomic weights, then it is possible to arrange them in a rather complicated table that demonstrates that certain properties of the elements repeat themselves periodically. If the table is arranged correctly, elements with similar properties fall into the same column. This is called the periodic table, and a workable version of it was first presented by the Russian chemist Dmitri Ivanovich Mendeleev (1834–1907) in 1869.

The periodic table was quite tentative at first because Mendeleev didn't know all of the elements. Many had not yet been discovered. In arranging the table so that similar elements were in the proper columns, Mendeleev was forced to leave gaps. He felt that these gaps represented undiscovered elements and, choosing three of those gaps, stated in 1871 that those undiscovered elements, once discovered, would have certain properties, which he described in detail. By 1885, all three elements were discovered and Mendeleev was proven precisely correct in each case. This offered very strong proof that the periodic table was a legitimate phenomenon, but no one could explain *why* it worked. (We will return to this later on.)

2

LIGHT

Particles and Waves

If we are prepared to admit that all matter is composed of atoms, then it is reasonable to ask if there is anything in the world that isn't matter and, therefore, isn't composed of atoms. The first possibility that might spring to mind is light.

It has always seemed obvious that light is immaterial. Solids and liquids can be touched; have mass, and therefore weight; and take up space. Gases cannot be felt in the same way that solids and liquids can, but a moving gas can be felt. We have all experienced high winds and we well know what a tornado can do. Then, too, air will take up room so that if an "empty" beaker (actually full of air) is plunged, open end down, into a tank of water, the water does not fill the beaker unless, somehow, the air is allowed to escape. In 1643, the Italian physicist Evangelista Torricelli (1608–1647) showed that air had weight and that this weight could support a column of mercury 76 centimeters (30 inches) high.

Light, however, has none of these properties. It cannot be felt, even though the heat it might produce can. It has never been found to have perceptible mass or weight, and it does not appear to take up space.

This doesn't mean that light was dismissed as unimportant because it was insubstantial. The first words of God, as given in the Bible are: "Let there be light." What's more, under the name of fire, it was the fourth of the ancient Earthly elements, on a par with the three material ones of air, water, and earth.

Sunlight was naturally considered to be light at its purest. It was white light, unchanging and eternal. If sunlight were made to pass through colored glass, it would pick up the color of the glass, but that would be an earthly impurity. Again, when objects burned on earth and gave off light, that light might be yellow, orange, or red. In some cases, if certain powders were cast into the fire, it might even burn green or blue. But again, these were earthly impurities that gave rise to color.

The one colored object that seemed to be divorced from anything earthly was the rainbow, which was sufficiently awe-inspiring to give rise to myths and legends. It was thought to be the bridge between heaven and earth, used by divine messengers. (The Greek messenger of the gods is given the name Iris, which is Greek for "rainbow.") It was also a divine guarantee that the world would never again be destroyed by flood, so that it appears at the end of rainstorms, indicating that God has remembered and stopped the rain.

In 1665, however, the English scientist Isaac Newton

27

(1642–1727) produced his own rainbow. In a darkened room, he allowed a beam of sunlight to enter through a hole in a shutter, and passed that beam through a three-dimensional triangular wedge of glass called a prism. The beam of light spread out and produced a band of colors on the white wall beyond, the colors being red, orange, yellow, green, blue, and violet, in that order—just the order in which they occur in the rainbow.

A rainbow, we now know, is caused by sunlight passing through the innumerable droplets of rain still in the air after a rainshower. These droplets have the same effect on light rays as a glass prism.

Apparently, then, sunlight is not "pure" light, after all. Its whiteness is merely the effect produced on the eye by a mixture of all of these colors. By having the light pass through a prism and then pass through another prism held in the reverse position, the separated colors will rejoin and form white light again.

In that these colors are thoroughly immaterial, Newton called the rainbow band a spectrum, from the Latin word for "ghost." Newton's spectrum created a problem, however. For the colors to be separated on passing through the prism, Newton believed, each one must have its ordinary straight-line path bent (refracted) as it passed into and out of the glass—each color bent to a different extent (red the least and violet the most), so that they were separated and seen each by itself when the beam hit the wall. What, then, could light be made of that would account for the separation of light into a spectrum?

Newton was an atomist and so it naturally occurred to him that light was made up of tiny particles, like the atoms of matter, except that the particles of light did not have mass. He had no clear notion, however, as to how the particles of colored light might differ among themselves, and why some should be refracted by a prism to a greater extent than others.

Furthermore, when two beams of light crossed each other, one remained unaffected by the other. If both consisted of particles, should not those particles collide and bounce off one another randomly so that the beam would grow fuzzy and spread outward after collision?

The Dutch physicist Christiaan Huygens (1629–1695) had an alternate suggestion. He thought light consisted of tiny waves. In 1678, he advanced arguments for showing that an entire series of waves might advance in what looked like a straight line, just as a beam of particles would, and that two beams, each made up of waves, would cross each other without either being, in the end, disturbed.

The trouble with the wave suggestion was that people thought of the types of waves produced in water, such as when a pebble is dropped into a still pond. As those water waves expand, they tend to move around an obstruction such as a piece of wood (diffraction) and join again on the other side. In that case, wouldn't light waves curve around an obstruction and cast no shadows, or at least fuzzy ones? Instead, as is well known, light casts sharp shadows if the light source is small and steady. Such sharp shadows are exactly what you would expect if light were a beam of minute particles, and this was considered a strong argument against waves.

It is interesting to note that the Italian physicist Francesco Maria Grimaldi (ca. 1618–1663) had noticed that a beam of light passing through two narrow openings, one behind the other, widened a little bit, indicating it had diffracted outward very slightly as it passed through the openings. His observation was published in 1665, two years after his death, but somehow it didn't attract attention. (In science, as in many other types of human endeavor, important discoveries or events sometimes get lost in the shuffle.)

Huygens, nevertheless, showed that light, if composed of waves, might well have waves of different lengths. Those portions of light with the longest waves would be least refracted. The shorter the waves, the greater the refraction. In this way, one could explain the spectrum, in that it might be that red had the longest waves and that orange, yellow, green, and blue were made up of successively shorter waves, while violet was made up of the shortest.

On the whole, as we look back on it, Huygens had the better of the argument, but Newton's reputation was growing rapidly (he was undoubtedly the greatest scientist who had ever lived) and it was hard to take up a position against him. (Scientists, in that they are as human as anyone else, are sometimes swayed by personalities as well as by logic.)

Throughout the 1700s, then, most scientists accepted the fact that light consisted of little particles. This might have helped the growth of atomism in connection with matter, and as atomism gained, that, in turn, strengthened the particle view of light.

In 1801, however, the English physicist Thomas Young (1773–1829) performed a crucial experiment. He let light fall upon a surface containing two closely adjacent slits. Each slit served as the source of a cone of light, and the two cones overlapped before falling on a screen.

If light were composed of particles, the overlapping region should receive particles from both slits and be brighter than the outlying regions that received particles from only one slit or the other. This was not so. What Young found was that the overlapping portions consisted of stripes—bright bands and dim bands alternating.

There seemed no way of explaining this phenomenon by the particle hypothesis. With waves, however, there were no problems. If the waves from one slit were in phase with those from the other slit, both keeping perfect step, then the ups and downs of one set of waves (or the ins and outs) would be reinforced by those of the other set, and the oscillation of the two combined would be stronger than of either separately. Brightness would increase.

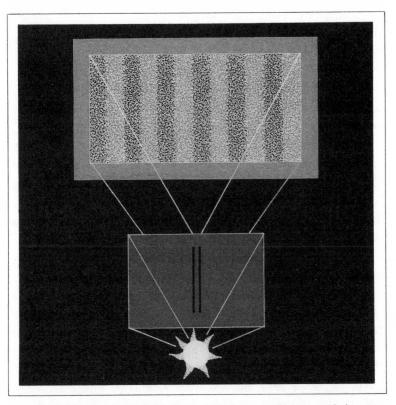

In 1801, Thomas Young let light fall on a surface containing two closely adjacent slits. The wedge of light from each slit fell on a screen and overlapped, resulting in a pattern of stripes—bright bands and dim bands alternating. There seemed no way of explaining this phenomenon by the particle hypothesis.

On the other hand, if the waves from one slit were out of phase with those from the other slit—if one set of waves went up while the other went down (or one went in while the other went out)—then the two waves would cancel each other, at least in part, and the two combined would be weaker than either separately. Brightness would decrease.

Young was able to show that, under the conditions of his experiment, the two sets of waves would be in phase in one region, out of phase in the next, in phase again in the next, and so on, alternately. The bright and dim bands

31

that were observed would be exactly what would be expected of waves.

Because one set of waves interferes with, and cancels, the other set in specific places, these bands are called interference patterns. Such interference patterns are observed when one set of waves on a calm water surface overlaps another. They are also observed when two beams of sound (known to consist of waves) intersect each other. The wave nature of light thus appeared to be demonstrated by Young's experiment (although, as we might expect, that didn't mean that those who believed in the particle view surrendered easily—because they didn't).

It was even possible, from the width of the interference bands, to calculate the length of a single wave of light (wavelength). It turned out that light waves had lengths in the neighborhood of $\frac{1}{20,000}$ of a centimeter (or $\frac{1}{50,000}$ of an inch). The wavelength of red light was a little longer than that, while the wavelength of violet light was a little shorter. This means that a ray of light an inch long will have, more or less, 50,000 waves, end to end, along the ray. It also means that about fifty atoms can be placed end to end along a single wavelength of light.

That explained why light cast sharp shadows despite being made of waves. Waves bend around obstacles only when the obstacles are not much longer than the wave in question. A wave would not bend around anything substantially longer than itself. Sound waves are very long and can move around most ordinary obstacles.

Almost anything we can easily see, however, is much, much longer than a light wave, so there's virtually no turning for them, and the shadows they cast are sharp. There is a very *slight* turning effect, however, and where the objects are quite small, the shadow's edge is inclined to be slightly fuzzy. That explains the diffraction effect that Grimaldi had discovered 130 years before Young's time.

The issue wasn't settled, however. People knew of two

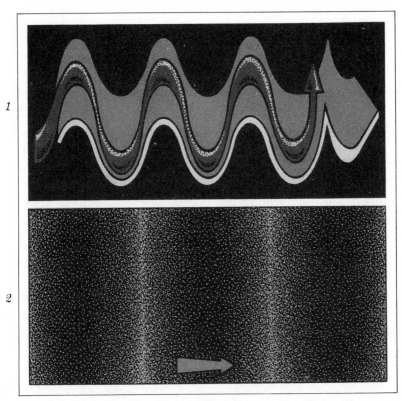

1. Water waves spread outward, and the particles of water move up and down in a direction at right angles to the direction in which the wave progresses. This type of wave is called a transverse wave.
2. Sound waves also spread outward, but the particles of air move in and out in a direction parallel to the direction in which the wave progresses. This type is called a longitudinal wave.

types of waves. There were water waves, in which the wave spread outward, but the particles of water moved up and down in a direction at right angles to the direction in which the wave progressed. This is called a transverse wave. There were also sound waves, in which the wave also spread outward, but the particles of air moved in and out, in a direction parallel to the direction in which the wave progressed. This is called a longitudinal wave.

Which of these two describes light waves? Huygens,

when he first elaborated the wave hypothesis, might have felt that light and sound, both being the cause of sense perceptions, should be similar in nature. Sound was known to be a longitudinal wave, so he suggested that light was a longitudinal wave, too. Young, when he demonstrated the wave nature of light, also thought so.

Earlier, in 1669, however, the Danish scholar Erasmus Bartholin (1625–1698) received a transparent crystal from Iceland, of a type now called Iceland spar. He noted that objects viewed through the crystal were seen double. He assumed that light passing through the crystal was refracted at two different angles, so that some emerged in one place and the rest in a slightly different place, producing a double image.

Bartholin could not explain why this should be, and neither could Newton or Huygens. The phenomenon was therefore pushed to one side as temporarily inexplicable. (Not everything can be explained at some particular stage of knowledge. The only sensible thing to do is to explain what you can and hope that, as knowledge advances, the time will come when the temporarily unexplainable can also be explained.)

In 1817, Young realized that double refraction could not be easily explained if light consisted of particles or of longitudinal waves. It could be explained quite easily, however, if light consisted of transverse waves.

The French physicist Augustin Jean Fresnel (1788–1827) adopted this point of view and worked out a careful theoretical study of light as transverse waves, one that explained all that was known about the behavior of light at that time. That settled it. For the next eighty years, physicists were quite satisfied that light was made up of tiny transverse waves and that that was the whole of the answer.

The Four Phenomena

It is a rare answer that is *completely* satisfactory, and this seems especially true in science, where every answer seems to uncover a more subtle question. If we grant that light exists in waves, as sound and a disturbed pond surface do, then there remains the problem that light waves travel easily through a vacuum, whereas sound waves and water waves do not.

Water waves exist because water molecules move up and down regularly. If water did not exist, then neither would water waves. Sound waves exist because air molecules (or the molecules of any medium through which sound travels) move in and out regularly. If air or any other medium does not exist, sound waves would not exist either.

In the case of light waves, however, what is it that is moving up and down? It can't be any type of ordinary matter, for light waves can pass through a vacuum where, apparently, there is no matter.

Newton had a similar problem when he worked out the law of universal gravitation in 1687. The Sun held the Earth in its gravitational grip across 150 million kilometers (93 million miles) of vacuum. How could the gravitational effect, whatever it was, travel across a vacuum?

Newton wondered if, perhaps, vacuum was not really *nothing* but consisted of a type of matter more subtle than ordinary matter and therefore not easily detectable. This vacuum matter came to be called ether, in homage to the "aether" Aristotle imagined as making up the heavenly bodies. The gravitational attraction pulled at the ether, and this pull was conducted from one bit of ether to the next until, finally, the Sun was pulling at the Earth.

Perhaps it was this ether (or another type) that waved up and down as light passed through. It had to fill all of space because we could see even the most distant stars. What's more, it had to be so fine and rarefied a type of matter that it did not in any way interfere with the passage of the Earth, or any other heavenly body, however light, as it progressed through space. Fresnel suggested that ether permeated the very body of the Earth and of all other heavenly bodies.

The particles of ether, however, when moved up, must experience a restoring force that moved them down, past the equilibrium point, then up again. The more rigid a medium, the more rapidly it vibrates up and down, and the more rapidly a wave progresses through it.

Light travels at a speed of 299,792 kilometers (186,290 miles) per second. This was first determined, very approximately, by the Danish astronomer Olaus Roemer (1644–1710) in 1676. To allow light to travel at such a speed, the ether must be more rigid than steel.

To have the vacuum made of something so fine that it allowed bodies to pass through it freely and without measurable interference, and at the same time so stiff as to be more rigid than steel, was rather puzzling, but scientists didn't seem to have any choice but to suppose that this was the case.

In addition to light and gravity, two other phenomena were known that could make themselves felt across a vacuum. They were electricity and magnetism. Both were first studied, according to tradition, by Thales. He studied a certain piece of iron ore, first found near the town of Magnesia on the eastern shore of the Aegean Sea. It had the property of attracting pieces of iron and he is supposed to have called it *ho magnetes lithos* ("the Magnesian rock"). Objects with the property of attracting iron have been called magnets ever since.

Thales also found that lumps of amber (a fossilized resin), if rubbed, attracted not iron particularly, but *any* light object. This difference in behavior meant the attraction was not that of magnetism. The Greek word for amber is *elektron* and, eventually, this phenomenon came to be called electricity as a consequence.

Sometime in the eleventh century, in China—but exactly where and by whom and under what circumstances is unknown—it was discovered that if a needle made of magnetic ore, or of steel that had been magnetized by being stroked by magnetic ore, was allowed to turn freely, it would align itself north and south. In addition, if the ends were marked in some way, it would be seen that the same end always turned north.

That end was called the magnetic north pole, and the other the magnetic south pole. In 1269, the French scholar Petrus Peregrinus (1240–?) experimented with such needles and found that the magnetic north pole of one would be attracted to the magnetic south pole of the other. On the other hand, the magnetic north poles of two magnetized needles would repel each other, as would the magnetic south poles of the two needles. In short, like magnetic poles repelled each other while unlike magnetic poles attracted each other.

In 1785, the French physicist Charles Augustin de Coulomb (1736–1806) measured the strength of the force by which a magnetic north pole attracted a magnetic south pole, or repelled another magnetic north pole. He found that the attraction or repulsion declined as the square of the distance (the inverse square law). That is, if you increased the distance to x times what it was before, the force between the poles became $1/x \times 1/x$, or $1/x^2$, what it was before. When Newton dealt with gravitation in 1687, he showed that the force of gravitational attraction followed what was to become the inverse square law.

Thus, the Moon is sixty times as far from the Earth's center as the Earth's surface is. The Earth's gravitational

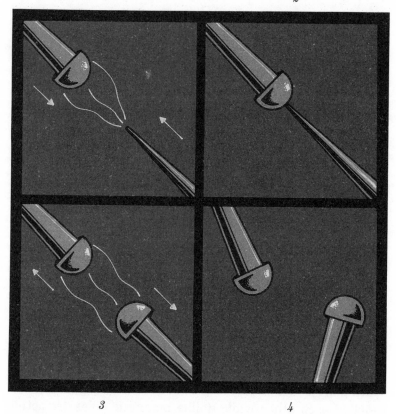

1. and 2. An unmagnetized iron needle will be attracted to either the north or the south pole of a magnet. Once magnetized, however, one end will be repelled, while the other will attract.
3. Unlike poles (N, S) attract.
4. Like poles (N, N or S, S) repel.

pull at the distance of the Moon is only ⅟₆₀ × ⅟₆₀, or ⅟₃₆₀₀ of what it is on the Earth's surface. Nevertheless, this pull is proportional to the product of the two masses involved, and the Earth and Moon are so massive that the Earth's gravitational pull is still large enough at the distance of the Moon to hold the Moon in orbit.

For that matter, the Sun can hold the Earth in orbit across a distance nearly 400 times that between the Earth

38

and the Moon. Indeed, huge clusters of galaxies, stretching across millions of light-years of space are held together by gravitational pulls.

Yet, as it turns out, the magnetic attraction between two magnetized needles is trillions of trillions of trillions of times as strong as the gravitational attraction between those same two magnetized needles. Why is it, then, that we are so aware of gravitational pulls and hardly at all aware of magnetic pulls? Why are astronomical bodies held together by gravitation, while we never hear of two bodies held together by some magnetic force?

The answer is that magnetism involves both an attraction and a repulsion, both of equal intensity. Gravitation involves *only* an attraction. There is no such thing as gravitational repulsion.

The world is full of magnets. As we shall see, every atom is a tiny magnet. The magnets of the Universe are turned every which way, however, and there is as much chance of repulsions here and there as attractions. On the whole, the two cancel each other and, by and large, we are left with a Universe in which there is not much magnetic attraction or repulsion overall.

Gravitation, however, involving only an attraction, simply piles up, so to speak. Although the effect of gravitational pull is so small as to be unnoticeable for ordinary objects, or even for mountains, by the time you have objects the size of the Earth, or the Sun, the gravitational pull is enormous.

Still, magnetism does play its part. Suppose you place a piece of stiff paper over a magnetized steel bar. Scatter some iron filings upon the paper and tap it. The tapping allows the filings to move and to take up some natural position with respect to the magnet. When this is done, the filings arrange themselves in a group of curved lines extending from one pole of the magnet to the other. Peregrinus had noticed this, and, in 1831, the English scientist Michael Faraday (1791–1867) considered the subject.

To Faraday it seemed that the influence of the magnet stretched out through space in all directions in a magnetic field that weakened with distance according to the inverse square law. Through the field one could draw a vast number of lines (magnetic lines of force) that marked out regions where the strength of the magnetic field was the same. Such lines were followed by the iron filings and were thus made visible.

That is why magnetic needles in a compass point north and south. The Earth itself is a magnet, and the needles line up with the magnetic lines of force that go from one of Earth's magnetic poles to the other. (Earth's magnetic poles are located in the far north and the far south, but are a considerable distance from the geographic poles of rotation.) A great many other facts about magnets can be explained by the concept of the magnetic field and its lines of force, and Faraday's notion has remained valid ever since. (There are also gravitational and electric fields, and lines of force there, too.)

What about electricity, by the way? The English physicist William Gilbert (ca. 1544–1603) extended Thales's work on electrified substances. He explained in a book he published in 1600 that substances other than amber would also attract light objects if rubbed. Gilbert called all such objects electrics.

In 1733, the French chemist Charles François de Cisternay Dufay (1698–1739) experimented with rods of glass and of resin, both of which could be "electrified," when rubbed, and made capable of attracting light objects. Both would then attract small bits of cork, which were, in turn, electrified.

A piece of cork electrified by glass would attract a piece of cork electrified by resin. Two pieces of cork, each electrified by glass, would, however, repel each other; as would two pieces of cork, each electrified by resin. Dufay concluded, therefore, that there were two types of electricity. Each repelled itself, but attracted the other, as in the case of the two types of magnetic poles.

The American scholar Benjamin Franklin (1706–1790) took this one step further. He suggested, in 1747, that there was but one type of electricity, which all matter contained a normal amount of and which was undetectable. If certain objects were rubbed, however, then some of the electricity was removed; while if other objects were rubbed, some was added. Those objects that had an excess might be considered positively charged, those with a deficiency were negatively charged.

In such a case a positively charged object would attract a negatively charged one because contact would allow the excess charge in the first to flow into the second and make up the deficiency there. The two would cancel each other and leave two uncharged objects behind. (This was actually observed by Franklin in his experiments.) On the other hand, two positively charged objects would repel each other as would two negatively charged objects because in neither case was there a chance of charge flowing from one to the other.

It only remained for Franklin to decide which of the two types of electrically charged bodies had the excess and which the deficiency. There was no way of telling at the time, so Franklin chose arbitrarily. He decided that rubbed glass had an excess and should be considered positive (+), while rubbed resins had a deficiency and should be considered negative (−).

Ever since, people working with electric currents have assumed that current flows from positive to negative. Unfortunately, Franklin had had a fifty-fifty chance of guessing right, and had lost. It was the rod of resin that actually

41

had the excess, so that the current really flows from negative to positive. That doesn't matter in electrical engineering, however. The results are the same whichever direction you imagine the current flowing, provided you stick by your decision and don't change your mind in midcourse.

Combining the Phenomena

There are, then, four phenomena that can make themselves felt across a vacuum: light, electricity, magnetism, and gravitation. All four might be pictured as making use of ether, but are they making use of the same ether, or does each one have an ether of its own? There was no way of telling, but sometimes light was pictured as waves in the luminiferous ether, from a Latin expression meaning "light carrying." Might there also turn out to be an "electriferous," a "magnetiferous," and a "gravitiferous" ether?

To be sure, the differences among the four were not equally great. Light did not seem either to attract or repel. Gravitation only attracted. Electricity and magnetism, however, each both attracted and repelled, and did so in much the same way with likes repelling and unlikes attracting. Of these last two, one seemed to arise out of the other.

In 1819, the Danish physicist Hans Christian Oersted (1777–1851) was lecturing on the electric current and, as a demonstration (it is not clear what he was trying to show), he brought a compass near a wire through which an electric current was flowing. To his own profound surprise, the compass needle reacted at once, pointing in a direction at right angles to the flow of current. When Oersted reversed the flow of current, the compass needle veered and pointed in the opposite direction, still at right angles to the current flow.

Oersted was the first to demonstrate an intimate connection between electricity and magnetism, but did not proceed in his investigations. Others, who heard of the demonstration, did, however, and at once.

In 1820, the French physicist Dominique François Arago (1786–1853) showed that a wire carrying an electric current acted as a magnet and could attract iron filings; something it would no longer do when the current ceased flowing. Because the wire was copper, this showed that magnetism was not necessarily an attribute of iron only, but might exist in any matter. Scientists began to speak of electromagnetism.

That same year, another French physicist, André Marie Ampère (1775–1836), showed that if two parallel wires had current flowing through each in the same direction, they attracted each other; if in opposite directions, they repelled each other.

If you twist a wire into a helix (the shape of a bedspring) and send a current through it, the current travels through each curve in the helix in the same direction. All of the curves attract one another and each sets up a magnetic field, every one reinforcing all of the others. The solenoid (coil of wire) then acts like a bar magnet with a north magnetic pole at one end and a south magnetic pole at the other.

In 1823, the British physicist William Sturgeon (1783–1850) wrapped wire about a U-shaped iron bar. The iron tended to intensify the magnetic field and, when the electric current was turned on, it became a surprisingly strong electromagnet.

In 1829, the American physicist Joseph Henry (1797–1878) used insulated wire (to prevent short-circuits), wrapping hundreds of turns of it about an iron bar, to produce an electromagnet that could lift phenomenal weights of iron when a current was passed through it.

Faraday then considered the reverse. If electricity could create magnetism, might not magnetism create electricity? He inserted an ordinary bar magnet into a helix of wire that was *not* connected to any battery that could start an electric current flowing in it. The magnet, nevertheless, experienced such a current when the magnet was pushed in, or pulled out. There was no current when the magnet was motionless at any point inside the helix. Apparently, the current flowed through the wire only when the wire cut across the magnetic lines of force, flowing in one direction as the magnet went in and in the other direction as it came out.

In 1831, Faraday worked out a system whereby a copper disc was turned between the poles of a magnet. An electric current was set up in the disc and flowed continually as long as the disc turned. It was an effort to keep it turning because it took work to push the disc across the magnetic lines of force. As long as this was done, however, by human or animal muscle, or by falling water, or by the force of steam produced by burning fuel, mechanical work was turned into electricity.

This time, it was Henry who reversed the situation. That same year, he invented the electric motor, in which the flow of electricity caused a wheel to turn.

All of these discoveries served to electrify the world (in a literal as well as a figurative sense) and to alter human society enormously. To scientists, however, the importance of these discoveries was that they increasingly demonstrated the close relationship between electricity and magnetism.

Indeed, there were those who began to think that there was a single electromagnetic field, one that, at times, showed its electrical face to the world, and, at times, its magnetic face. This reached its climax with the work of Maxwell. Between 1864 and 1873, he worked out the mathematical implications of Faraday's notions of fields and lines of force, and of the apparent connection of electric and magnetic fields. Maxwell ended by devising four comparatively simple equations (simple to mathematicians, at any rate) that described all known electrical and magnetic behavior. They have been known ever since as Maxwell's equations.

Maxwell's equations (whose validity is confirmed by all observations made since) show that electric fields and magnetic fields cannot exist separately. There is, indeed, only a combined electromagnetic field with an electric component and a magnetic component at right angles to each other.

If electric behavior and magnetic behavior were similar in all respects, the four equations would be symmetrical; they would exist in two mirror-image pairs. In one respect, however, the two phenomena do not match each other. In electrical phenomena, positive charges and negative charges can exist independently of each other. An object can be either positively charged or negatively charged. In magnetic phenomena, on the other hand, the magnetic poles do not exist separately. Every object that shows magnetic properties has a north magnetic pole at one location and a south magnetic pole at another location. If a long magnetized needle, with a north magnetic pole at one end and a south magnetic pole at the other, is broken in the middle, the poles are *not* isolated. The end with the north magnetic pole instantly develops a south magnetic pole at the break, while the end with the south magnetic pole develops a north magnetic pole at the break.

Maxwell included this fact in his equations, which introduced a note of asymmetry. This has always bothered

45

scientists, in whom there is a strong drive for simplicity and symmetry. This "flaw" in Maxwell's equations is something we'll return to later.

Maxwell showed that from his equations you can demonstrate that an oscillating electric field will produce, inevitably, an oscillating magnetic field, which will in turn produce another oscillating electric field, and so on indefinitely. This is the equivalent of an electromagnetic radiation moving outward, in wave form, at a constant speed. The speed of this radiation can be calculated by taking the ratio of certain units expressing magnetic phenomena to other units expressing electrical phenomena. This ratio works out to nearly 300,000 kilometers (186,290 miles) per second, which is the speed of light.

This could not be a coincidence. Light, it appeared, was an electromagnetic radiation. Maxwell's equations thus served to unify three of the four phenomena known to pass through a vacuum: electricity, magnetism, and light.

Only gravitation remained outside this unification. It seemed to have nothing to do with the unified three. Albert Einstein, in 1916, worked out his general theory of relativity, which improved on Newton's concept of gravitation. In Einstein's interpretation of gravity, which is now widely accepted as essentially correct, there should be gravitational radiation in the form of waves, analogous to electromagnetic radiation. Such gravitational waves, however, are much more subtle and feeble, and much more difficult to detect, than are electromagnetic waves. Despite some false alarms, they have not yet been detected at this moment of writing, although virtually no scientist in the field doubts that they exist.

Extending the Spectrum

Maxwell's equations set no limitations on the period of oscillations of the field. There could be one oscillation per second or less, so that each wave would be 300,000 kilometers long, or more. There could also be a decillion oscillations per second or more, so that each wave would be a trillionth of a trillionth of a centimeter long. And there could be anything in between.

Light waves, however, represent only a tiny fraction of these possibilities. The longest wavelengths of visible light are 0.0007 millimeters long, and the shortest wavelengths of visible light are just about half this length. Does this mean there is electromagnetic radiation we don't see?

Through most of human history, the question as to whether light existed that could not be seen would have been considered a contradiction in terms. Light, by definition, was something that could be seen.

The German-British astronomer William Herschel (1738–1822) was, in 1800, the first to show this was not a contradiction after all. At that time, it was thought that the light and heat one obtained from the Sun might be two separate phenomena. Herschel wondered if heat might be spread out in a spectrum just as light was.

Instead, then, of studying the spectrum by eye, which noted only the light, Herschel studied it by thermometer, which measured the heat. He placed the thermometer at various places in the spectrum and noted the temperature. He expected that the temperature would be highest in the middle of the spectrum and that it would fall off at either end.

That did not happen. The temperature rose steadily as one progressed away from the violet, and reached its highest point at the extreme red. Astonished, Herschel won-

47

dered what would happen if he placed the thermometer bulb *beyond* the red. He found, to his even greater astonishment, that the temperature rose to a higher figure there than anywhere in the visible spectrum. Herschel thought he had detected heat waves.

In a few years, however, the wave theory of light was established and a better interpretation became possible. Sunlight has a range of wavelengths that are spread out by a prism. Our retina reacts to wavelengths of light within certain limits, but sunlight has some waves that are longer than that of the visible red, and is therefore to be found beyond the red end of the spectrum. Our retina won't respond to such long waves, so we don't see them, but they are there, anyway. They are called infrared rays, the prefix coming from a Latin word meaning "below," for you might view the spectrum as going from violet on the top to red on the bottom.

All light, when it strikes the skin, is either reflected or absorbed. When absorbed, its energy speeds up the motion of the molecules in our skin and this makes itself felt as heat. The longer the wavelength, the deeper it penetrates the skin and the more easily absorbed it is. Hence, although we can't see the infrared, we can feel it as heat, and the thermometer, for similar reasons, can record it as such.

It would help, of course, if it could be shown that infrared rays were actually made up of waves like those of light, but with longer wavelengths. One might allow two beams of infrared rays to overlap and produce interference fringes, but no one would be able to see them. Perhaps they could be detected by thermometer, with the temperature going up each time the instrument passed through a "brighter" area, and going down each time it passed through a "dimmer" one.

In 1830, the Italian physicist Leopoldo Nobili (1784–

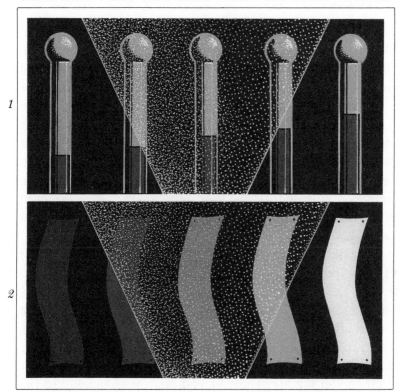

1. *Infrared Light. In 1800, when William Herschel placed his thermometer in the dark area beyond the red end of the spectrum, he was surprised to record the highest temperature.*
2. *Ultraviolet Light. In 1770, Karl Wilhelm Scheele found that paper soaked in silver nitrate solution darkened most quickly when exposed to violet light. In 1801, John Wilhelm Ritter exposed the paper in the dark area beyond the violet and the paper darkened even more rapidly.*

1835) invented a thermometer that would do the job. One of his co-workers was the Italian physicist Macedonio Melloni (1798–1854). Because glass would absorb a great deal of the infrared rays, Melloni made use of prisms formed of rock salt, which is transparent to infrared rays. As a result, interference fringes were set up and Nobili's thermometer showed that they existed. By 1850, Melloni had demonstrated that infrared rays showed all of the properties of light without exception—except that they could not be seen with the naked eye.

What about the other end of the spectrum, where violet light deepens into darkness? That story began in 1614, when the Italian chemist Angelo Sala (1576–1637) noticed that silver nitrate, a perfectly white compound, darkened on exposure to sunlight. This happens, we now know, because light contains energy and can force apart the molecule of silver nitrate, producing finely divided silver, which appears black.

About 1770, the Swedish chemist Karl Wilhelm Scheele (1742–1786) went into the subject in more detail, making use of the solar spectrum, which wasn't known in Sala's time. He soaked thin strips of white paper in solutions of silver nitrate, let them dry, and placed them in various parts of the spectrum. He found that the strips of paper darkened least quickly in the red, more quickly as one went farther and farther from the red, and most quickly in the violet. This happens (as we now know, for reasons that will be explained later) because light increases in energy as one goes from red to violet.

Once Herschel discovered infrared rays in 1810, however, it occurred to the German chemist Johann Wilhelm Ritter (1776–1810) to check the other end of the spectrum. In 1801, he soaked strips of paper in silver nitrate solution and repeated Scheele's experiment except that he placed strips of paper *beyond* the violet, where no light was visible. As he suspected they might, the strips of paper darkened in this lightless region even more rapidly than they would in violet light. This represented the discovery of ultraviolet rays, where the prefix is from the Latin for "beyond."

Infrared and ultraviolet radiation existed just at the borders of the visible spectrum. Maxwell's equations made it seem that there could be radiation far beyond the borders. If such radiation could be found, then Maxwell's equations would be supported very strongly, for without them no one would have suspected such radiation might exist.

In 1888, the German physicist Heinrich Rudolf Hertz (1857–1894) made use of a rectangular wire, with a gap in it, as a detecting device. He set up an oscillating electric current in his laboratory. As the electric current oscillated, moving first this way, then that, it should emit electromagnetic radiation, with the radiation wave moving up while the current is going one way and then down when it is moving the other way. Such an electromagnetic wave should have a very long wavelength because even if the oscillating electric current changes direction every small fraction of a second, light can move quite far between changes.

Hertz's rectangular wire would gain an electric current if the electromagnetic wave crossed it, and there would be a spark across the gap. Hertz got his spark. In addition, as he moved his rectangular wire here and there in the room he got a spark where the wave was very high or very low, but no spark where it was in between. In this way, he could map the wave and determine its length.

Hertz had discovered what came to be called radio waves, which lay far beyond the infrared radiation and could have wavelengths of anywhere from centimeters to kilometers.

No one questioned Maxwell's equations after that. If there was a luminiferous ether, it carried electricity and magnetism also. If there was another ether, it existed only for gravitation.

In 1895, by the way, electromagnetic radiation was discovered far beyond the ultraviolet, with wavelengths exceedingly small; but we will get to that later, after we consider a few other matters.

Dividing Energy

Electricity, magnetism, light, and gravitation are all forms of energy, where energy is anything that can be made to do work. These forms of energy certainly seem different from one another, but one can be turned into another. As we have already seen, electricity can be turned into magnetism, and vice versa, and a vibrating electromagnetic field can produce light. Gravitation can cause water to fall, with the falling water turning a turbine that can force a conductor through magnetic lines of force to produce electricity. Interconversions of energy and work represent the field of thermodynamics.

Such conversions are never completely efficient. Some energy is always lost in the process. The lost energy does not, however, disappear, but makes its appearance as heat, which is still another form of energy. If heat is taken into account, then no energy is ever totally lost, nor does any energy ever appear out of nowhere. In other words, the total amount of energy in the Universe seems to be constant.

This is the law of conservation of energy, or the first law of thermodynamics, which was finally placed in compelling terms in 1847 by the German physicist Hermann Ludwig Ferdinand von Helmholtz (1821–1894).

In a way, heat is the most fundamental form of energy, for any other form of energy can be converted *completely* into heat, while heat cannot be converted completely into nonheat energy. For this reason, heat is the most convenient phenomenon through which to study thermodynamics; a word, by the way, which is from the Greek for "movement of heat."

Heat had been closely studied by scientists ever since the first truly practical steam engine had been invented, in 1769, by the British engineer James Watt (1736–1819).

52

Once the law of conservation of energy was understood, the study of heat became even more intense.

After the advent of the steam engine, there were two theories of the nature of heat. Some scientists thought of it as a type of subtle fluid that could travel from one piece of matter to another. Others thought of heat as a form of motion, of atoms and molecules moving or vibrating.

The latter suggestion, or the kinetic theory of heat (where kinetic is from a Greek word for "motion"), was finally established in the 1860s as the correct one when Maxwell and the Austrian physicist Ludwig Eduard Boltzmann (1844–1906) worked it out mathematically. They showed that everything that was known about heat could be interpreted satisfactorily by dealing with atoms and molecules that were moving or vibrating. As in the case of gases, the average speed (or, better, "velocity") of motion or vibration of the atoms and molecules making up *anything* is the measure of its temperature if the mass of the atoms and molecules is also taken into account. The total kinetic energy (which takes into account both mass and velocity) of all of those moving particles is the total heat of the substance.

Naturally, then, the colder an object gets, the slower the motion of its atoms and molecules. If it gets cold enough, the kinetic energy of the particles reaches a minimum. It can then get no colder, and the temperature is at absolute zero. This notion was first proposed and made clear in 1848 by the British mathematician William Thomson (1824–1907), better known by his later title of Lord Kelvin. The number of Celsius degrees above absolute zero is the absolute temperature of a substance. If absolute zero is equal to $-273.15°$ C, $0°$ C is equal to $273.15°$ K (for Kelvin) or $273.15°$ A (for absolute).

Any body at a temperature higher than that of its surroundings tends to lose heat as electromagnetic radiation. The higher the temperature, the more intense the radiation. In 1879, the Austrian physicist Joseph Stefan (1835–1893) worked this out exactly. He showed that the total radiation increased as the fourth power of the absolute temperature. Thus, if the absolute temperature was increased two times, say from 300° K to 600° K (that is, from 27° C to 327° C), then the total radiation would be increased $2 \times 2 \times 2 \times 2$, or 16 times.

Formerly, about 1860, the German physicist Gustav Robert Kirchhoff (1824–1887) had established the fact that any substance at a temperature lower than that of its surroundings would absorb light of particular wavelengths, and would then emit those same wavelengths when its temperature rose above that of its surroundings. It follows that if a substance absorbs *all* wavelengths of light (a "black body," in that it reflects none of them), it will emit all wavelengths when heated.

No object actually absorbs all wavelengths of light, in the usual sense of the word, but an object with a small hole in it does so after a fashion. Any radiation that finds its way into the hole is not likely to find its way out again and is finally absorbed in the interior. Therefore, when such an object is heated, black-body radiation—all of the wavelengths—should come pouring out of the hole.

This notion was first advanced by the German physicist Wilhelm Wien (1864–1928) in the 1890s. When he studied such black-body radiation, he found that a wide range of wavelengths was emitted, as was to be expected, and that the very long and very short wavelengths were low in quantity, with a peak somewhere in between. As the temperature rose, Wien found that the peak moved steadily in the direction of shorter wavelength. He announced this in 1895.

Stefan's law and Wien's law fit our experience. Suppose an object is at a temperature a little higher than that of our own body. If we put our hands near that object, we can feel a little warmth radiating from it. As the temperature of the object rises, the radiation becomes more noticeable and the peak radiation is at a shorter wavelength. A kettle of boiling water will deliver considerable warmth if our hand is placed near it. If temperature is raised still higher, an object will eventually give off perceptible radiation at wavelengths short enough to be recognized by our retina as light. We first see red light because that is the light with the longest wavelength, and is the first to be emitted. The object is then red-hot. Naturally, most of the radiation is still in the infrared, but the tiny fraction that comes off in the visible portion of the spectrum is what we notice.

As the object continues to rise in temperature, it glows more and more brightly. The color changes, too, as more and more of the shorter-wave light is emitted. As the object continues to grow still hotter, it becomes even brighter and the color undergoes another change as more, and shorter, wavelengths of light are emitted. The glow becomes more orange, and then yellow. Eventually, when something is as hot as the Sun's surface, it is white-hot, and the peak of the radiation is actually in the visible light region. If it grows still hotter, it becomes blue-white, and, eventually, although it is brighter than ever (assuming we can look at it without destroying our eyes in the same instant), the peak is in the ultraviolet.

This heat/light progression created a problem for nineteenth-century scientists because it was difficult to make sense out of the pattern of black-body radiation. Toward the end of the 1890s, the British physicist John William Strutt, Lord Rayleigh (1842–1919), assumed that every wavelength had an equal chance of being radiated in black-body radiation. On that assumption, he worked out an equation that showed quite well how the radiation would increase in intensity as one went from very long wavelengths to shorter wavelengths. This equation, however, didn't provide for a peak wavelength, to be followed by a decline, as one approached still shorter wavelengths.

Instead, the equation implied that the intensity would continue going up without limit as the wavelengths got shorter. This meant that any body should radiate chiefly in the short wavelengths, getting rid of all of its heat in a blast of violet, ultraviolet, and beyond. This is sometimes called the violet catastrophe. But the violet catastrophe does not take place, so there must be something wrong with Rayleigh's reasoning. Wien himself worked out an equation that would fit the distribution of short wavelengths of black-body radiation, but it wouldn't fit the long wavelengths. It seemed as though physicists could explain either half of the radiation range, but not the whole.

The problem was taken up by the German physicist Max Karl Ernst Ludwig Planck (1858–1947). He thought there might be something wrong with Rayleigh's assumption that every wavelength had an equal chance of being radiated in black-body radiation. What if the shorter the wavelength, the less the chance of its being radiated?

One way of making this seem plausible is to suppose that energy is not continuous and can't be broken up into smaller and smaller pieces forever. (Until Planck's time, the continuity of energy had been taken for granted by physicists. No one had wondered if energy, like matter, might consist of tiny particles that couldn't be divided further.)

Planck assumed that the fundamental bit of energy was larger and larger as the wavelength grew smaller and smaller. This meant that for a given temperature, the radiation would rise in intensity as wavelengths grew shorter, just as the Rayleigh equation indicated. Eventually, though, for wavelengths shorter still, the mounting size of the energy unit would increase the difficulty of getting

56

enough energy into one place in order to radiate it. There would be a peak, and as the wavelengths continued to decrease, the radiation would actually decline.

As the temperature went up and the heat grew more intense, it would be easier to radiate the larger energy units and the peak would move in the direction of shorter wavelengths, just as Wien's law would require. In short, the use of the energy units that Planck postulated completely solved the problem of black-body radiation.

Planck called these energy units quanta (quantum in the singular), which is a Latin word meaning "how much?" What counted, after all, in the answer to the black-body radiation puzzle was how much energy there is in the quanta of different wavelengths of radiation.

Planck advanced his quantum theory, and the equation it made possible for black-body radiation (which agreed with the actual observations both for long wavelengths and short wavelengths), in 1900. This theory proved so important—far more important than Planck, at the time, could possibly imagine—that all of physics prior to 1900 is called classical physics, and all of physics after 1900 is called modern physics. For his work on black-body radiation, Wien received a Nobel prize in 1911, and Planck received one in 1918.

3

ELECTRONS

Dividing Electricity

Early experiments on electricity dealt with objects that carried rather small electric charges. In 1746, however, the Dutch physicist Pieter van Musschenbroek (1692–1761), working at the University of Leyden, invented something called the Leyden jar, in which a great deal of electric charge could be pumped.

The greater the charge, the greater the pressure for discharge. If a Leyden jar is touched to some object, the electricity flows into the object and the jar is discharged. (If touched to a human being, that human can receive a flow of electricity that will surely be painful.)

If a Leyden jar carries a sufficiently large electric

charge, direct contact need not be made. Under such conditions, if the Leyden jar merely approaches an object that will discharge it, the electric charge can force its way from the Leyden jar to the other object through the intervening air.

The result is a flash of light and a crackle. The light is not the electricity itself; rather, the electricity, whatever it is, heats the air as it passes through, and the air grows momentarily hot enough to radiate light. The heat also expands the air and, after the discharge is complete, the expanded air contracts again and the crackling sound results.

Some people saw a similarity between the light and crackle of a discharging Leyden jar and the lightning and thunder in the clouds during a storm. Could lightning and thunder be the discharge of a gigantic Leyden jar arrangement in the clouds?

In 1752, Benjamin Franklin proved this was so by flying a kite in a thunderstorm, leading the charge of the lightning down a cord and into an uncharged Leyden jar. The resultant charged Leyden jar showed that electricity from the sky had the same properties as electricity produced on Earth.

But what about the electricity itself that hides inside the charged body, or inside the light produced by heated air? One answer would be to discharge electricity through a vacuum in order to see what the bare electricity looked like. As early as 1706, an English physicist, working with charged objects far less intense than a Leyden jar, managed to get a discharge across an evacuated vessel, obtaining light as he did so.

In those days, however, evacuating a vessel was still only imperfectly possible. A remnant of air would be left inside, and that would be enough to glow as a result of the passage of the electricity. It was not the electricity itself. In order to get bare electricity, two things were needful.

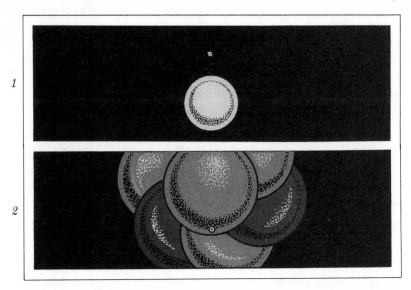

1. If you could see an atomic nucleus from the perspective of an orbiting electron, the nucleus might appear as a tiny dot at a great distance.

2. Even though they have charges of equal but opposite magnitude, an electron has only $1/1,837$ the mass of a single proton.

One was a vessel that was so well evacuated that the traces of air inside would not be enough to interfere with the electricity. The second was a way of forcing electricity in quantity through a good vacuum. A Leyden jar might do here, but its discharge lasted for only a moment. Was there a way of keeping the electricity going for a considerable period of time?

The second problem was solved in 1800 by the Italian physicist Alessandro Giuseppe Volta (1745–1827). He showed that electricity could be produced when two different metals were both dipped into a salt solution. This was accompanied by a chemical reaction, and as long as the chemical reaction proceeded, electricity continued to be produced. If some of the electricity was drawn off through a wire, the electricity would flow through the wire as long as it was being produced by the chemical reactions.

As a result, it became possible to have an electric current, instead of merely a stationary electric charge. To produce a large electric current, Volta made use of a series of two-metal combinations with salt water between. Any of a series of similar objects is called a battery. Volta had invented an electric battery.

As soon as Volta announced his discovery, scientists began constructing bigger and better batteries, and within a generation Faraday worked out a method for producing an electric current, much more cheaply, by burning fuel. There was then no problem of setting up electric currents with enough force to move across a good vacuum—if a good vacuum could be supplied.

The supplier was the German inventor Johann Heinrich Wilhelm Geissler (1814–1879), who, in 1855, invented an air pump that was a marked improvement on all of the pumps used previously. Instead of using mechanical devices involving moving parts, Geissler used only rising and falling levels of mercury. The mercury trapped a bit of air with each change in level and removed it. It was a slow process, but by the time this mercury air pump was done, over 99.9 percent of the air in a vessel had been removed.

Geissler, who was an expert glass blower, blew vessels that had two pieces of metal sealed into opposite ends, and these vessels were evacuated. Such vessels were named Geissler tubes by Geissler's friend and co-worker, the German physicist Julius Plücker (1801–1868). Plücker connected the two pieces of metal sealed in the tube to opposite poles of an electricity-generating device. One of these metal pieces, therefore, became positively charged and was called the anode, while the other became negatively charged and was called the cathode.

These words were first used by Michael Faraday. The positively charged anode is from Greek words meaning "upper way," and the negatively charged cathode from Greek words meaning "lower way." Since Benjamin Franklin's time, electricity was thought to flow from positive to negative; that is, from anode (upper) to cathode (lower), like water flowing from an upper level to a lower one.

Plücker forced electricity through the vacuum of a Geissler tube and now there was simply not enough air to make a visible glow—but there was a glow anyway. It was a greenish glow in the neighborhood of the cathode, *always* the cathode. Plücker reported his observations in 1858, and this was the first indication that Franklin might have made a wrong guess, and that electricity flowed not from anode to cathode but from cathode to anode.

Could that greenish glow represent the bare electric current itself? Plücker wasn't sure. He thought it might be pieces of the metal broken off and glowing, or that it might be the tiny wisps of gas still left in the vessel.

The German physicist Eugen Goldstein (1850–1930) studied the phenomenon carefully and found that it didn't matter what gas was in the vessel before it was evacuated. It also didn't matter what metal the anode and cathode were made of. The only thing that was the same in all cases was the electric current, so Goldstein maintained that the glow *was* associated with the current itself. In 1876, he called the vacuum-crossing material cathode rays.

This name implied that the current was emitted by the cathode and traveled to the anode. Indeed, the glass glowed on the anode side of the tube as though the cathode rays were striking and energizing it.

In 1869, the German physicist Johann Wilhelm Hittorf (1824–1914), who had been a student of Plücker, showed that if a solid object were sealed into the tube in front of the cathode, there would be a shadow of the object against the glow at the anode end. Clearly, something was traveling from the cathode, and some of it was stopped by the solid object.

The British physicist William Crookes (1832–1919) de-

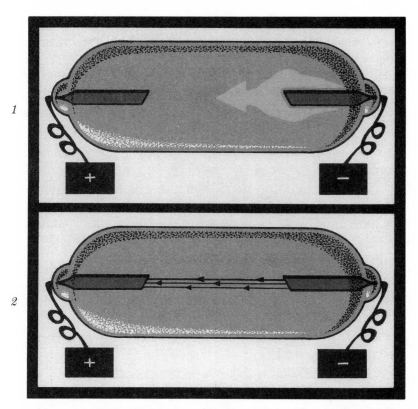

1. *In 1858, Julius Plücker reported a greenish glow around the negatively charged cathode in the near vacuum of a Geissler tube. This was the first indication that Ben Franklin's guess about the direction of current flow might be wrong.*
2. *In 1876, Eugen Goldstein maintained that the glow was associated with the current itself. He called the vacuum-crossing material* cathode rays. *Current flows from the cathode to the anode.*

vised still better devices for making a vacuum and, in 1878, produced a Crookes tube, in which the remaining air was only $\frac{1}{75,000}$ that in Geissler tubes. (All such tubes came to be grouped together as cathode-ray tubes.) The cathode rays now showed up more clearly, and Crookes could demonstrate that they moved in straight lines, and that they could even be made to turn a little wheel.

But what was it that flowed out of the cathodes? Did the cathode rays consist of particles or of waves? Both possibilities found support among scientists in a reprise of the type of argument about light between Newton and Huygens. The arguments in favor of particles for the cathode rays were somewhat the same as those in the case of light, the chief one being that the cathode rays cast sharp shadows.

The fact that the particle notion had been decisively defeated in the case of light made a number of scientists hesitate, however, to be caught on what might again be the losing side. (Generals are often accused of always being prepared for the previous war. Scientists are human, and they remember past battles, too, and sometimes have a tendency to bring old experiences to bear on new phenomena.)

The strongest voice in favor of the cathode rays being waves was that of Hertz, the discoverer of radio waves. In 1892, he showed that cathode rays could penetrate thin films of metals. It didn't seem to him that particles ought to be able to do so, but waves might, of course, because if the metal films were thin enough, even light waves could penetrate them.

Hertz's pupil Philipp Eduard Anton von Lenard (1862–1947) even prepared a cathode-ray tube with a thin, aluminum "window." The cathode rays were able to spray through the window and emerge into the open air. If the cathode rays were waves of very short length, they would travel in straight lines and cast sharp shadows, just as light waves do. For a while in the early 1890s, the notion of cathode rays as waves was therefore riding high.

And yet if the cathode rays were emerging from the negatively charged cathode, might they not be carrying a negative electric charge? If so, that might well indicate that cathode rays were not waves, for no wave known at that time carried any electric charge, however small. And if cathode rays did carry an electric charge, they ought to be affected by an electric field.

In 1883, Hertz tested the hypothesis. He passed cathode rays between two parallel metal plates, one of which was positively charged and one negatively charged. If the cathode rays were charged, they should veer out of their straight-line paths, but they did not. Hertz concluded that they were uncharged, and that, too, was a point in favor of waves.

What Hertz didn't realize, however, was that the cathode rays were traveling far more quickly than he expected, so that they got past the plates before they had a chance to veer perceptibly. This could have been taken care of if the plates were sufficiently highly charged, but they weren't. The combination of speedy cathode rays and weakly charged plates made the deviation from straight-line path imperceptible, so that Hertz's conclusion wasn't valid. (Scientific experiments are not always the last word. A particular experiment, however honestly and intelligently conducted, can yield the wrong answer for any of a variety of reasons. That is why it is important that experiments be checked by other scientists, using other instruments, other conditions, and, if possible, other ideas.)

Thus, in 1895, Perrin (who, in the following decade, was to demonstrate the reality of atoms) showed that cathode rays could impart a large negative charge to a cylinder on which they fell. It was hard to see how cathode rays could carry a negative charge from the cathode to the cylinder without themselves possessing the negative charge while they traveled. This seriously weakened Hertz's finding.

Then the British physicist Joseph John Thomson (1856–1940) decided to try repeating Hertz's experiment with electrically charged plates. Thomson had the advantage of knowing how rapidly cathode rays moved. In 1894, he had estimated that they moved at a speed of 200 kilometers (125 miles) per second. What's more, he had a more effectively evacuated vacuum tube than Hertz had had, and he used electrically charged plates with a considerably stronger charge.

In 1897, Thomson allowed cathode rays to speed between his charged plates and found that the electric field induced a distinct curvature away from the negatively charged plate and toward the positively charged one. That convinced him, and it convinced physicists, generally, that cathode rays consisted of speeding cathode-ray particles, each carrying a negative electric charge.

The verdict here was the reverse of that in the case of light. Where light was concerned, waves had won out over particles. With cathode rays, particles had won out over waves. (As we shall see, however, neither victory was absolute. It often happens in science that the choice between alternatives is not as clearcut as it might at first seem.)

Cathode-Ray Particles

The degree to which a charged particle is deflected by an electric field depends on three things: the size of the electric charge carried by the particle, the speed at which the particle travels, and the mass of the particle. The deflection of a charged particle by a *magnetic* field depends on the same three factors, but in a different fashion from the deflection by an electric field. If Thomson measured both types of deflection, it would be possible from the two measurements, taken together, to work out the ratio of the charge to the mass of the particles. Given that, if you knew the

66

size of the electric charge of the particles, you could then work out their mass.

The electric charge was also not entirely impossible to work out. Faraday had made extensive studies of the way in which electrical currents induced chemical reactions, and he had established the laws of electrochemistry in 1832. From these laws, and from careful measurements of the amount of electricity required to bring about the deposition of a known mass of metal from a solution of its compounds, it was possible to calculate the size of the electric charge required to deposit a single atom of the metal.

There seemed no great risk in deciding that the electric charge involved in the chemical change of a single atom might be the smallest electric charge that could exist. It was reasonable, therefore, to suppose that a cathode-ray particle would carry this smallest electric charge. In other words, a cathode-ray particle would be to electricity what an atom is to matter—or, as was about to be discovered, what a quantum is to energy.

Taking this assumption and the measured deflections of cathode rays by an electric field and by a magnetic field of known strength, Thomson could calculate the mass of a single cathode-ray particle, and did so. He was awarded a Nobel prize in 1906 for this accomplishment.

The results were surprising. As far as atoms of matter are concerned, the smallest atom known in Thomson's day (and in our own) was the hydrogen atom. Indeed, we are now quite certain that an ordinary hydrogen atom is the smallest atom that can possibly exist. The cathode-ray particle, however, turned out to have a mass far smaller than that of the hydrogen atom. It has a mass only $\frac{1}{1837}$ that of the smallest atoms.

For a century, scientists had been quite certain atoms were the smallest things that could exist, and that the smallest atom would therefore be the smallest *anything*

67

that had mass. Now that thought was shattered; or, at least, it had to be modified, but the modification might not have to be very great. It was possible to argue, after Thomson's experiments, that atoms were still the smallest pieces of *matter* that could exist. Electricity, it might be said, was not matter, but a form of energy that was much more subtle than matter. It should not be surprising, from that point of view, that these cathode-ray particles, which might be viewed as "atoms of electricity," were much smaller than atoms of matter.

It was the smallness of the cathode-ray particle that might account for the fact that an electric current could flow through matter, or that cathode-ray particles could themselves pass through thin films of metals. The passage of these particles through metal had been taken as strong evidence that they could not be particles, but at the time of the first discovery of such passage, there had been no idea of how *small* those particles were. (Experiments can mislead even the best scientists if some key bit of knowledge is missing.)

Because the cathode-ray particle is far smaller than any atom, it is termed a subatomic particle. It was the first subatomic particle to be discovered, and was to be the first of a flood of them that would completely change our minds about the structure of matter. Their discovery increased our knowledge, revolutionized our technology, and utterly changed our way of life. (The topic of technology and our way of life is outside the scope of this book, but the fact is worth mentioning. No matter how ivory-towerish scientific discoveries might seem, there is always a good chance that they will affect us in many crucial ways.)

What does one call a cathode-ray particle? Naming something does not increase our knowledge concerning it, but it makes it easier to refer to it and to discuss it. In 1891, the Irish physicist George Johnstone Stoney (1826–

1911) had suggested that the minimum electric charge that one could deduce from Faraday's laws be called an electron. Thomson liked the name and applied it to the particle rather than to the electric charge it carried. The name has stuck and become very familiar even to the nonscientific public (think of all the electronic devices, such as television sets and record players, that we deal with). We might say, then, that Thomson discovered the electron in 1897.

X Rays

In the previous chapter I mentioned that electromagnetic radiation lying far beyond the ultraviolet in the shortwave direction was eventually discovered. I did not go into detail then, but the time has now come when we can discuss it.

In the 1890s, the German physicist Wilhelm Konrad Roentgen (1845–1923) was working on cathode rays in his own unique way. He did not concern himself, as Hertz and Thomson had, with their nature, but with their effect on certain chemicals. Cathode rays, impinging on those chemicals, caused them to luminesce. That is, the chemicals gained energy from the cathode rays and then lost that energy again in the form of the radiation of visible light.

One of the chemicals that luminesced upon the impingement of cathode rays was a compound called barium platinocyanide. Roentgen had sheets of paper coated with that compound in his laboratory.

The luminescence was quite faint and, in order to observe it as well as possible, Roentgen darkened the room and enclosed the experimental apparatus within sheets of black cardboard. He could then peer into an enclosure that was totally dark, and when he turned on the electric current, the cathode rays would pass along the tube, penetrate

69

the thin, far wall, fall upon a chemically coated paper, and initiate luminescence that he could see and study.

On November 5, 1895, Roentgen turned on the current and, as he did so, a dim flash of light that was *not* inside the apparatus caught the corner of his eye. He looked up and there, quite a distance from the apparatus, was one of the sheets, covered with barium platinocyanide, luminescing briskly.

Roentgen turned off the current and the coated paper darkened. He turned it on and the coated paper glowed again. He took the paper into another room and pulled down the blinds in order to darken the room. When he turned on the cathode-ray tube, the coated paper glowed in this room.

Roentgen decided that the cathode-ray tube was producing radiation that was not cathode rays—a radiation that could penetrate cardboard, and even the wall between two rooms, as cathode rays could not. He published his first report on this new radiation on December 28, 1895, and because he had no idea of the nature of the radiation, he called it X rays. The name has clung to the radiation ever since. For this discovery, Roentgen received a Nobel prize in 1901, the first year in which such prizes were given out.

Now the same problem and uncertainty arose over X rays that had previously arisen over light and over cathode rays. Some physicists thought X rays were streams of particles, some thought they were waves. Of those who thought they were waves, some (like Roentgen himself) thought they were longitudinal waves, like sound waves. Others thought they were transverse waves, like light waves. If they were transverse waves, they might be a type of electromagnetic radiation with wavelengths far shorter than ultraviolet, just as the recently discovered radio waves had wavelengths far longer than the infrared.

The problem was how to decide among the alternatives. Light had been shown to be waves because it displayed interference. In order to demonstrate the interference, light had been passed through two closely spaced slots. The interference could be made more pronounced by using diffraction gratings, glass plates on which very finely spaced parallel scratches were made. Light passing through the intervals between the scratches produced clearly visible interference phenomena, allowing wavelengths to be measured with great precision.

The shorter the wavelength, however, the finer the spacing between the scratches must be. Diffraction gratings wouldn't work on X rays if they were transverse waves with extremely short wavelengths. It then occurred to the German physicist Max Theodor Felix von Laue (1879–1960) that it was not necessary to try to manufacture a diffraction grating with impossibly closely spaced scratches. Nature had already done the job.

Crystals consist of atoms and molecules of a substance arranged in an all but endless even array. This can be inferred from the shapes of crystals and from their tendency to break in certain planes in such a way as to retain their shapes. It is as though they break "with the grain" along a plane that lies between two adjacent layers of atoms of molecules. Why should not X rays, Laue reasoned, penetrate the crystals between the layers? A crystal might serve as a diffraction grating with scratches no wider apart than the layers of its atoms, and this might show interference effects for X rays.

If X rays were to go through an object in which atoms and molecules were scattered in random disorder, the X rays would be scattered this way and that, in random fashion. There would be a uniform shadowing effect, darkest at the center and growing lighter as one moved outward in all directions.

If X rays were to go through a crystal with orderly layers of atoms and molecules, the X rays' diffraction pat-

71

terns would be set up and the photographic plate would show distinct spots of light and shadow forming a symmetrical pattern about the center.

In 1912, Laue tried the experiment of passing X rays through a crystal of zinc sulfide. It worked perfectly, the X rays behaving exactly as they would be expected to if they were very short transverse waves. That settled the issue and, in 1914, Laue received a Nobel prize for his work.

The British physicist William Henry Bragg (1862–1942), together with his son, William Lawrence Bragg (1890–1971), a physics student at Cambridge, saw that X-ray diffraction could be used to determine the actual wavelength of X rays if the distance between the layers of atoms in the crystal diffracting the rays were known. This they accomplished in 1913, showing that the wavelength of X rays was anywhere from $\frac{1}{50}$ to $\frac{1}{50,000}$ the wavelength of visible light. For this they shared a Nobel prize in 1915.

Electrons and Atoms

It is clear, when one stops to think of it, that electrons might exist in matter. Suppose we consider the early studies of electricity, when one simply built up an electric charge by rubbing a glass rod or a piece of amber. Might this not be because electrons travel from the object being rubbed to the object doing the rubbing, or vice versa? Any substance that gets extra electrons forced into it will accumulate a negative charge, and any substance that loses some of its electrons will accumulate a positive charge. And if so, the electrons have to be in the matter to begin with if they are going to be transferred one way or the other.

Again, an electric current might consist of electrons moving through the material in which the current exists.

Thus, in a cathode-ray tube, when the electric current reaches the cathode, electrons accumulate there (giving it a negative charge, which is what makes it a cathode) and are forced into the vacuum as a stream of cathode-ray particles.

The electrical *impulse* travels at the speed of light, so that if you have wires strung from a telephone in New York to one in Los Angeles, a voice can modulate the electric flow in New York, which will then reproduce the voice in Los Angeles about 1/60 of a second later. The electrons themselves, however, bumping from atom to atom, travel much more slowly.

This is analogous to what happens when you flick a checker against a long line of similar checkers. As the checker you flick strikes the first in the long line, the last, at the other end of the line, flies away almost at once. The checkers in the middle barely move, but the impulse of compression and expansion moves along the line of checkers at the speed of sound and ejects the last checker.

Still, though it seemed quite likely that electrons might well exist in matter, it was somehow taken for granted that these particles of electricity existed quite apart from, and independently of, atoms, which were pictured as featureless and indivisible.

Information gathered from chemical experiments during the 1800s certainly made it seem that atoms were indivisible, but that they were featureless was a mere assumption. Nevertheless, scientists are human, and in science, as in other facets of human thought, an assumption that has been held long enough sometimes takes on the force of cosmic law. People forget that it is only an assumption and find it difficult to consider the possibility that it might be wrong.

In this connection, consider the manner in which an electric current can pass through some solutions and not others. This phenomenon was first studied systematically by Michael Faraday.

Thus, a solution of table salt (sodium chloride) will conduct electricity, as Volta found when he constructed the first electric battery. Sodium chloride is therefore an electrolyte. An electric current will *not* pass through a solution of sugar; therefore, sugar is a nonelectrolyte.

From his experiments, Faraday decided that something in the solution carried negative charges in one direction and positive charges in the other. He didn't know exactly what it was that carried the charge, but he could give it a name. He called the charge carriers ions, from a Greek word meaning "wanderers."

In the 1880s, a young Swedish chemical student, Svante August Arrhenius (1859–1927), tackled the problem in a novel way. Pure water has a certain fixed freezing point at 0° C. Water that has a nonelectrolyte dissolved in it (say, sugar) freezes at slightly below 0° C. The more sugar dissolved in the water, the lower the freezing point. In fact, the lowering of the freezing point is proportional to the number of molecules of sugar dissolved in it. This holds true for other nonelectrolytes, too. The same number of molecules of any nonelectrolyte in solution will lower the freezing point by the same amount.

The situation is different with electrolytes. If sodium chloride is dissolved in water, then the freezing point is lowered just twice as much as it ought to be, considering the number of molecules in solution. Why should that be?

Sodium chloride has a molecule made up of one atom of sodium (Na) and one of chlorine (Cl), so that its formula is NaCl. When sodium chloride is dissolved in water, Arrhenius suggested, it breaks up, or dissociates, into those two atoms, Na and Cl. For every molecule of NaCl outside of solution, you have two half molecules, Na and Cl, so to speak, in solution. There would be twice as many particles in solution as was thought, and there would be twice the lowering of the freezing point. (Molecules made up of more than two atoms might break up into three or even four parts and produce three, or even four, times the expected lowering of the freezing point.)

A molecule of ordinary sugar has a molecule consisting of 12 carbon atoms, 22 hydrogen atoms, and 11 oxygen atoms, or 45 atoms altogether. When it dissolves in water, however, it does not dissociate, but remains in full molecular form. Therefore, there are only the expected number of molecules in solution and the freezing point is lowered only by the expected amount.

When sodium chloride dissociates, however, it can't possibly break up into ordinary sodium and chlorine atoms. The properties of sodium and chlorine atoms are known, and are not to be found in a salt solution. Something must happen that makes the sodium and chlorine of dissociated sodium chloride different from ordinary sodium and chlorine.

To Arrhenius, it seemed the answer was that each dissociated fragment of the sodium chloride molecule carried an electric charge and that they were the ions that Faraday had spoken of. From the results of the experiment involving an electric current passing through a sodium chloride solution, it was easy to argue that each sodium particle formed through dissociation carried a positive charge and was a sodium ion that could be symbolized as Na^+, while each chlorine particle carried a negative charge and was a chloride ion, symbolized as Cl^-. It was because electrolytes tended to dissociate into such charged fragments that they *were* electrolytes and could conduct an electric current.

Sodium ions and chloride ions had properties far different from uncharged sodium atoms and chlorine atoms. That is why a salt solution is a mild substance, while sodium and chlorine, themselves, are both dangerous to life. Non-

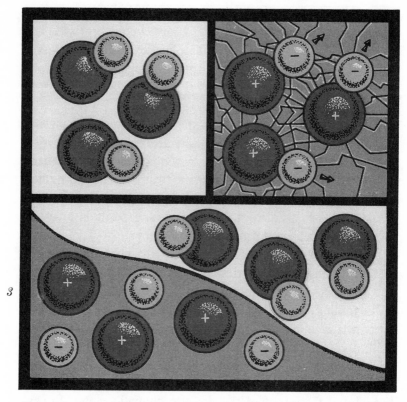

1. and 2. When ordinary table salt, sodium chloride (NaCl), dissolves in water, it dissociates into positively charged sodium ions, Na+, and negatively charged chloride ions, Cl-.
3. Atoms in and out of solution.

electrolytes, such as sugar, not being dissociated, have no charged fragments that can carry an electric charge, and therefore do not conduct an electric current.

In 1884, Arrhenius prepared his theory of ionic dissociation as a thesis for the degree of Doctor of Philosophy. The examining committee greeted the thesis with coldness, for they were not prepared to accept any theory that spoke of atoms that carried an electric charge. How could atoms carry an electric charge when atoms were featureless and

76

incapable of modification? (They were helpless in the grip of the assumption.)

The committee could not actually reject the thesis because it was perfectly argued, and because it explained so many things that couldn't be explained any other way. Nevertheless, they passed it with the lowest possible passing grade.

When, thirteen years later, J. J. Thomson discovered the electron, it suddenly became obvious that atoms might, just possibly, carry one or two excess electrons or lose one or two of the normal quantity they might contain. With each passing year, there were new discoveries that made that possibility seem more certain, and in 1903, Arrhenius received a Nobel prize for the very thesis that, nineteen years before, had barely earned a passing grade.

Of course, it wasn't entirely satisfactory to deduce the presence of electrons in atoms merely from the behavior of electrolytes. Was there any way of observing electrons in atoms directly? Could one, for instance, knock electrons out of atoms and detect them?

In 1887, when Hertz was experimenting with the detecting device with which he was to prove the existence of radio waves the following year, a spark appeared across the gap of his detecting device whenever electricity jumped the gap. He observed something curious though, for a spark appeared more easily when light shone upon the gap.

Apparently, light had some effect on electric discharge, so this came to be called the photoelectric effect, the prefix photo- coming from the Greek word for "light."

In the very next year, 1888, another German physicist, Wilhelm Hallwachs (1859–1922), found that the photoelectric effect did not treat the two types of electric charge equally. A piece of the metal zinc, carrying a negative charge, lost that charge when it was exposed to ultraviolet rays. That same piece of zinc, carrying a positive charge, was not affected at all by ultraviolet radiation, and retained its charge. There was no ready explanation for this until Thomson discovered the electron and it began to appear that electrons might exist in matter.

In that case, a spark was formed across a gap because electrons were forced out of one of the metal points at the gap. If light somehow caused electrons to be ejected, the spark would form more easily. Again, zinc that had a negative charge would be carrying an excess supply of electrons, and if light caused those electrons to be ejected, the zinc would lose its charge. Zinc that had a positive charge would have a deficiency of electrons, and because light could not be expected to supply electrons to make up that deficiency the positive charge would remain unaffected.

At least, that was the easy explanation of the early observations of the photoelectric effect. It is, however, an advisable caution on the part of scientists not to rush toward the easy explanation too precipitously. Sometimes, one can fall into a trap that way (as when one decides that cathode rays can't consist of particles because they pass through thin films of metal).

Thus, just because electrons are knocked out of matter does not mean they necessarily exist in matter to begin with. Einstein, in 1905, showed as part of his special theory of relativity that mass was a form of energy. Mass could be turned into energy, and energy could be turned into mass.

Light contained energy. Might it be, then, that light-energy on striking metal under certain conditions would be converted into a tiny fragment of mass—an electron—that would carry off a bit of the negative charge possessed by the metal? In this way, electrons would appear that had never been part of the metal.

Einstein's theory, however, did not merely state that mass and energy were interchangeable. It presented a simple equation that showed just how much mass would be converted into how much energy and vice versa. It turns out that even a small quantity of mass could be turned into a great deal of energy; and, conversely, that it took a great deal of energy to form even a small amount of mass.

The electron is a particularly small bit of mass, but even so, the quantity of energy it would take to form it is simply not present in ultraviolet rays, as was soon to be determined. The photoelectric effect cannot, therefore, be the result of the creation of electrons out of energy; it must be the result of the ejection of electrons already present in the metal atoms.

It takes far less energy to eject an electron that already exists than to form one from scratch. In this case, then, the simpler explanation turned out to be correct (and it is a pleasant thing, indeed, that this sometimes happens in science).

Of course, it was still possible that what emerged from the metal might not be electrons. They might be some other type of particle carrying a negative charge. In 1899, however, Thomson applied magnetic and electric fields to the emerging particles and found that they had the same mass as electrons and the same negative charge. With those two properties matching, it seemed clear that photoelectric particles were electrons, and there has been nothing to disturb that view since.

Electrons and Quanta

Philipp Lenard studied the photoelectric effect in 1902 and was able to show that the electrons ejected from various metals always matched each other in properties. In other words, although there were many different atoms, they were all associated with but one type of electron. This was a hopeful bit of information considering that scientists love simplicity.

On the other hand, Lenard found that not all light was equal when it came to inducing a photoelectric effect. It often happened that red light did not produce the ejection of electrons, and that making the light more intense didn't help. No electrons would appear no matter how intense the light was.

However, if one exposed a particular metal to light of shorter and shorter wavelengths, there came a point at which electrons began to be ejected. The wavelength at which this happens is called the threshold value.

At the threshold value, the electrons that are ejected move at a very slow speed, as though the light has just barely enough energy to eject them and no more. If the light at the threshold value is made more intense, more electrons are ejected—but they still move at a very slow speed.

If the metal is exposed to light with wavelengths smaller and still smaller than that of the threshold value, the electrons are ejected with greater and greater speed. The speed of the electrons depends on the wavelength, while the number of electrons ejected depends on the intensity of the light. Different metals have different threshold values, as though some metals hold electrons more loosely than other metals.

Lenard couldn't explain this, and neither could J. J. Thomson when he tried. Ordinary nineteenth-century physics didn't work. When the solution did come, it came by way of quantum theory, which had been devised by Planck five years earlier.

Planck had supposed that electromagnetic radiation came in quanta of a certain size. The shorter the wavelength, the larger the energy content of the quantum.

It is also true that the shorter the wavelength, the greater the number of waves the radiation can produce in one second. The number of waves of radiation per second is called the frequency. The shorter the wavelength, then, the higher the frequency. We can therefore say that the size of a quantum is proportional to its frequency.

Until 1905, the notion of quanta had only been used in connection with black-body radiation. Might it not be a mathematical trick that explained that one phenomenon and nothing more? Did quanta *really* exist?

Einstein, whose theoretical work in 1905 made it possible to show, a few years later, that atoms really existed, tackled this new question concerning reality in that same year.

Einstein was the first to take the quantum theory seriously, and to consider it more than a mere convenience in solving the one problem of black-body radiation. He was willing to suppose that energy came in quanta at all times and under all conditions, so that problems that involved energy, other than black-body radiation, must also take quanta into consideration.

This meant that radiation existed in quanta form when it struck matter. It struck as quanta and, if absorbed, was absorbed as quanta. At any one moment in any one place, an entire quantum is absorbed; nothing more, nothing less.

If light that strikes is of long wavelength and low frequency, the quanta are of low energy. Such a quantum, when absorbed, simply might not contain enough energy to break an electron loose from a particular atom. In such a case, the quantum is absorbed as heat, and the electron might vibrate faster but it doesn't break away. Given enough quanta of this sort, a substance might absorb enough heat to melt, but at no moment in time is enough heat absorbed by any single atom to shake an electron loose.

As the wavelength decreases and the frequency increases, the quantum contains more energy and, at the threshold value, there is just enough energy to break an electron loose. There is no excess energy to appear as energy of motion, so the electron moves very slowly.

With still shorter wavelengths and still more energetic quanta, there is enough additional energy to eject an electron with considerable speed. The shorter the wavelength and the more energetic the quanta, the faster that motion.

Depending on the nature of the atom, electrons are held more tightly or more loosely, to begin with, and would then require larger quanta, or smaller quanta, to bring about an ejection. The threshold value will therefore be different for each element.

The quantum theory neatly explained all of the observed facts about the photoelectric effect, and this was very impressive. When a theory that has been worked out to explain one phenomenon turns out to explain another phenomenon, apparently unrelated to the first, it becomes very tempting to accept the theory as representing reality. (Here you see an example of the use of a theory; it explains widely different categories of observations. Without quantum theory, no one could see the connection between black-body radiation and the photoelectric effect—to say nothing of many other phenomena.) It was for his work in this connection that Einstein received a Nobel prize in 1921.

Waves and Particles

If light occurs in quanta, and if each quantum goes speeding separately through space, the quantum behaves, in that way, like a particle. The quantum even received a name in its particle aspect. Because of the electron, most particles have received an -on ending, and, in 1928, the American

82

physicist Arthur Holly Compton (1892–1962) named such a speeding quantum a photon, from the Greek word for "light."

It was fitting that Compton invented the name, for in 1923, he showed that radiation *did* act as particles, not just by being separate pieces of something, but by *behaving* as particles did. The shorter the wavelength and the more energetic the quanta, the more likely it was that they would demonstrate properties usually considered characteristic of particles rather than of waves.

Compton studied the manner in which X rays were scattered by crystals and found that some X rays, in the process of being scattered, increased their wavelengths. This meant that some of the energy of the X-ray quantum was lost to an electron in the crystal. Compton thought that the effect might be particulate in nature, like that of one billiard ball hitting another, with one losing energy and the other gaining it. He found, when he worked out a mathematical relationship that accurately described what had happened, that this seemed to be so in actual fact. This is now called the Compton effect.

It turned out, then, that both Newton and Huygens had hold of part of the truth two and a half centuries before. Light consisted of something that was both wave and particle. This can be confusing. In the ordinary world around us, there are waves, such as water waves; and there are particles, such as sand particles; and there are no confusions about it. Waves are waves and particles are particles.

The point is that light does not resemble the ordinary objects around us, and can't be forced into categories defined according to the same rules. Light, when studied in certain ways, shows interference phenomena, as water waves do. Studied in other ways, however, they show energy transfers, as colliding billiard balls do. No observation, however, can show light acting both as a wave and a particle

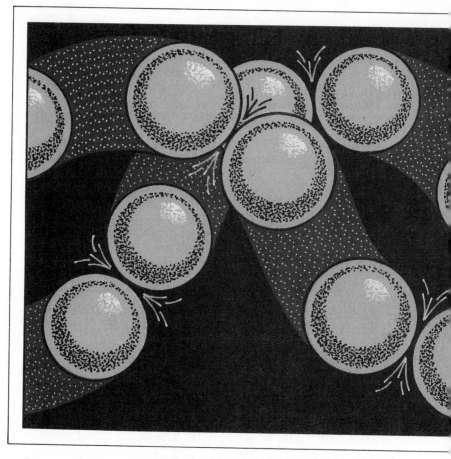

In 1923, Arthur Holly Compton found that X rays behaved as particles, losing energy when they were scattered by crystals. An X ray behaved like one billiard ball hitting another, with one losing energy

simultaneously. You can study light as either one or the other, never both at once.

This is not really such a mystery. Imagine you are looking at an empty ice-cream cone from the side, so that the wide part is at the top and the point is at the bottom. The outline is that of a triangle. Imagine next that you are looking at it with the wide opening facing you directly, and the point directed away from you. Now the outline is that

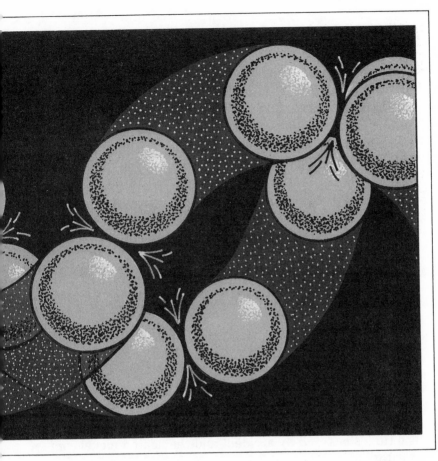

and the other gaining it. The shorter the wavelength, and the more energetic the quanta, the more likely that it would demonstrate properties considered characteristic of particles rather than waves.

of a circle. If those are the only two ways in which you are allowed to view the cone, then you can see it as *either* a circle *or* a triangle, but you can never see it as both simultaneously.

You might ask what the *real* two-dimensional outline of the cone was, but the answer would have to be "It depends on how you look at it." In the same way, you might ask whether light was really a wave, or really a particle, and you would have to answer, "It depends on the particular way in which you are observing it."

One great side effect of the particle nature of light was that it made the luminiferous ether unnecessary. After a century of existence in scientists' minds, during which time the necessity of explaining it wrought more and more confusion, it disappeared as though it had never been—and, in fact, it never *had* been.

If something that had seemed to be a wave turned out to have a particle aspect, might it not be that something that had seemed to be a particle would turn out to have a wave aspect?

The French physicist Louis Victor de Broglie (1892–1987) suggested, in 1924, that this might be so. He made use of Einstein's equation relating mass to energy and Planck's equation relating the size of a quantum to frequency to show that every particle ought also to act like a wave of a certain length.

In 1925, the American physicist Clinton Joseph Davisson (1881–1958) was studying the reflection of electrons from a metallic nickel target enclosed in a vacuum tube. The tube shattered by accident and the heated nickel combined with oxygen from the air, rusting the surface of the target. To remove the film, Davisson had to heat the nickel for an extended period. Once this was done, it turned out that the electron-reflecting properties of the nickel surface had changed. The surface had consisted of many tiny crystals before the accident, but of just a few large ones afterward.

Davisson, who knew of de Broglie's suggestion, thought it would be useful to go still further and to prepare a nickel surface made of a single crystal. This might be able to show any wave aspect an electron might have. He aimed a stream of electrons at the single-crystal surface and found that electrons were not only reflected, but were diffracted and showed interference phenomena. Electrons *did* have a wave aspect.

Also in 1925, the British physicist George Paget Thomson (1892–1975), only son of J. J. Thomson, was forcing fast electrons through very thin gold films and he, too, noticed diffraction effects. As a result, de Broglie received a Nobel prize in 1929 for working out the theory of electron waves, while Davisson and G. P. Thomson shared one in 1937 for demonstrating it. Electron waves are *not* electromagnetic waves, by the way, but are "matter waves."

Physicists are now convinced that, indeed, everything has both a particle and a wave aspect, but not necessarily in equal measure. The more massive a particle is, the more prominent the particle aspect is and the more difficult it is to observe the wave aspect. A billiard ball (or the Earth itself) has a wave aspect, but this has so short a wavelength that it is quite likely it might never be observed. We know it is there in theory, but that is all. Even a grain of sand has a wave aspect too subtle to be observed, in all likelihood. An electron, however, has such a small mass that its wave aspect can be observed easily, once the proper experiment is carried out.

In the same way, the less energetic a wave is, the more prominent its wave aspect, and the more difficult it is to observe its particle aspect. A water wave is so weak (if only a single molecule of water is taken into account) that it is not at all likely its particle aspect can be observed. The same is true of a sound wave, although physicists speak of the particle aspect of sound waves as phonons, from the Greek word for "sound."

Even electromagnetic radiation is hard to observe in its particle aspect when the quanta are very small, as in radio waves. It is only when the quanta grow large and the wavelengths tiny, as in X rays, that the particle aspect can be easily observed.

87

Einstein showed that the gravitational field should radiate waves just as an electromagnetic field does. The gravitational field is so much less intense than the electromagnetic that gravitational waves are exceedingly weak, and their particle aspect almost hopelessly undetectable. Nevertheless, physicists speak of gravitational waves as consisting of speeding gravitons.

It is only because in the ordinary world about us particles are so massive and waves so lacking in energy that we think of the two phenomena as mutually exclusive. In the world of the atom and of subatomic particles, this exclusivity disappears.

Sometimes science is said to produce paradoxical results and to go against common sense. It is important to remember that common sense is often based on the very limited observations we make in the world about us. To go against common sense sometimes means that we are taking a broader and more accurate view of the Universe. (Remember that "common sense" once told us that the Earth is flat and that the Sun moves around it.)

4

NUCLEI

Probing the Atom

Once scientists began to suspect that electrons might be associated with atoms, a problem arose. Electrons carried a negative electric charge, but atoms were electrically neutral. That meant there had to be positive electric charges located somewhere in the atom that served to neutralize the charges of the electron.

If this were so, then if electrons were removed from an atom, what remained would carry a positive charge. If electrons were added to an atom, the atom plus the excess electrons would carry a negative charge. This would account for the positive and negative ions of Faraday and Arrhenius.

J. J. Thomson, in 1898, was the first to suggest an atomic structure that took electric charge into account. He maintained that the atom was a tiny, featureless sphere, as had been assumed for a century, but that it carried a positive electric charge. In that positively charged atom, enough electrons were embedded (like raisins in a cake) to neutralize the charge.

Thomson's notion of atomic structure still left the atom a solid affair, and if many atoms were lined up in contact, right and left, up and down, inward and outward, then the solid that was formed in this way must be exactly what the name indicated—*solid*.

Yet that could not be so. Lenard had noted in 1903 that the speeding electrons making up the cathode rays slipped through thin films of metal, which seemed to imply that an atom must consist, at least to some degree, of empty space. Lenard suggested that an atom consisted of a cloud of small particles; some of them being electrons and some of them being positively charged particles of similar size. A positive and a negative particle would revolve about each other, making up a pair that was, overall, neutral. A large cluster of such pairs might make up an atom, but between and within these pairs there would be empty space through which a small object such as a speeding electron could easily pass.

But if that were so, then an atom ought to lose either type of particle with equal ease. If exposure to light caused the ejection of negatively charged electrons from metal, why were not positively charged particles also ejected—at least now and then? Again, if speeding electrons left a cathode under the force of an electric current, why did not speeding positively charged particles emerge from the anode? Clearly, if the positively charged particles existed, they would have to be quite different in nature from electrons. The positive particles would have to be, for some reason, much less mobile than electrons.

In 1904, the Japanese physicist Hantaro Nagaoka (1865–1950) suggested that the positive portion of the atom did not take up the whole of the volume, as Thomson had suggested, and that it did not take up as much volume as the electrons, as Lenard had suggested. Nagaoka offered a compromise. Nagaoka believed that the positively charged portion of the atom was located in the atomic center, and was smaller than the atom as a whole. It was surrounded by circling electrons, held by electromagnetic attraction, as the planets circled the Sun, held by gravitational attraction.

Nagaoka's suggestion provided a neutral atom under ordinary conditions, allowed for the production of positive and negative ions, and left empty space for speeding electrons to pass through. In addition, it explained why electrons were so easily removed from atoms, while positively charged particles were not. The electrons, after all, were on the outskirts of the atom, while the positively charged portion was in the protected center.

Still, none of these suggestions really won acceptance. They were all speculative and uncompelling. What was needed was direct evidence concerning the internal structure of the atom. Such direct evidence did not seem as though it would be easy to come by. After all, how could anyone possibly probe the interior of so small an object as an atom? And yet, even as Thomson, Lenard, and Nagaoka were advancing their suggestions, such an atom-probing device already existed. The story of its discovery goes as follows.

As soon as X rays were discovered by Roentgen, other physicists hastened to study the new radiation, and many wondered if it could be found elsewhere, in places where it hadn't been noticed only because no one had thought of looking for it there.

The French physicist Antoine Henri Becquerel (1852–1908) was particularly interested in fluorescent compounds, substances that absorb sunlight (or other energetic radiation) and then give up the energy by emitting light of just a few restricted wavelengths. Fluorescence is very similar to phosphorescence, except that fluorescent substances cease to give off light as soon as they are no longer exposed to energetic radiation, while phosphorescent substances continue to give off light for a period of time after exposure ceases.

Becquerel wondered if fluorescent substances emitted X rays along with visible light. In order to test this, he planned to wrap photographic plates in black paper and put the package in sunlight with a crystal of a fluorescent chemical upon it. The sunlight would not penetrate the black paper, and neither would any fluorescent light the crystal gave off. If, however, the crystal gave off X rays, these would penetrate the black paper and fog the photographic film.

The crystals he used were of potassium uranyl sulfate, a well-known fluorescent material. Each molecule of that compound contained one atom of the metal uranium.

On February 25, 1896, Becquerel performed his experiment and, sure enough, the photographic film was fogged. He decided that the crystal was indeed giving off X rays, and he prepared to repeat the experiment with new film in order to make sure. However, there followed several cloudy days. Becquerel put the photographic film, with its black paper wrapping and the crystal on top, in a drawer and waited for sunlight.

By the first of March Becquerel was restless. To give himself something to do, he decided to develop the film just to make sure that nothing was getting through the dark paper in the absence of fluorescence. To his amazement, something *was* getting through, and a lot of it. The plate was strongly fogged. The crystals must be giving off radiation that did not depend on sunlight and did not involve fluorescence. Forgetting the Sun, Becquerel began to study the radiation instead.

He quickly realized that the radiation given off by the potassium uranyl sulfate originated in the uranium atom, for other compounds containing uranium atoms gave off similar radiation, even when they were not fluorescent. In 1898, the Polish-French physicist Marie Curie (1867–1934) showed that another metal, thorium, also gave off radiation. She termed such behavior on the part of uranium and thorium radioactivity. Both Becquerel and Curie suspected that more than one type of radiation was involved.

In 1899, the New Zealand-born physicist Ernest Rutherford (1871–1937) studied the manner in which radioactive radiations penetrated sheets of aluminum. He found that some of the radiation could be stopped by $\frac{1}{500}$ of a centimeter of aluminum, while the rest required a considerably thicker sheet to be stopped. Rutherford called the first type of radiation alpha rays, from the first letter of the Greek alphabet, and the second type beta rays, from the second letter. A third type of radiation, which was the most penetrating of all, was discovered in 1900 by the French physicist Paul Ulrich Villard (1860–1934), and was called gamma rays, from the third letter of the Greek alphabet.

It was not long before these various radiations were quantified. The beta rays were deflected by a magnetic field in such a way that it was clear they consisted of negatively charged particles. In 1900, Becquerel determined the mass and the size of the charge of these particles, and it turned out that beta rays, like cathode rays, were made up of speeding electrons. A speeding electron is, therefore, sometimes called a beta particle.

Gamma rays were not deflected by a magnetic field, and this made it appear that they did not carry an electric charge. Rutherford suspected that gamma rays might be electromagnetic in nature, and passed some through a crystal. The existence of a diffraction pattern showed that they were very much like X rays, except that they possessed even shorter wavelengths.

As for alpha rays, they were deflected by a magnet in such a way as to show that they consisted of positively charged particles. Might these be the positively charged particles that, along with electrons, Lenard thought made up atoms?

No. Lenard had imagined that the positively charged particles were, rather, like electrons in their properties except for the nature of the charge. The alpha particles, however, were very different from electrons in other ways than their electric charge. In 1906, Rutherford showed that the alpha particle was much more massive than an electron. We now know that it is about 7,344 times as massive as an electron.

As soon as Rutherford found the alpha particles to be particularly massive, it seemed to him that they would be the very thing with which to probe the atom. A stream of alpha particles striking a thin film of metal would penetrate it, and the manner of its penetration might yield useful information.

Rutherford placed a piece of radioactive substance in a lead box that had a hole in it. The radiations could not penetrate the lead, but a thin stream of radiation would emerge from the hole, and, traveling outward, would strike a thin film of gold. Behind the gold was a photographic plate, which would be fogged by any alpha particles that passed through the gold.

The gold sheet was so thin as to be semitransparent, but, just the same, so tiny are atoms that that same sheet was about 20,000 gold atoms thick. Even so, the alpha particles smashed through as though those 20,000 atoms simply weren't there. They fogged the photographic plate in precisely the spot that would have been fogged if the gold film had not been there.

Yet not entirely. Rutherford noticed that a few alpha particles were deflected. There was a faint haze of clouding around the dark central spot on the photographic plate. The haze faded off quickly with distance, but didn't entirely disappear. About 1 in 8,000 alpha particles were deflected ninety degrees or more. In fact, an occasional alpha particle seemed to hit something and bounce directly backward.

To explain this, Rutherford advanced his idea of atomic structure in 1911. The atom, he said, had almost all its mass concentrated in a small, positively charged body at its very center. At the outskirts of the atom, spread over a volume that took up almost all of the atom, were nothing but electrons. This was something like Nagaoka's atom, except that the positively charged body at the center of the atom was much smaller and more massive in Rutherford's atom.

Furthermore, Rutherford had experimental observations, which Nagaoka did not. The alpha particles penetrated the electron portion of the atom as though it were empty because the alpha particle was so much more massive than the electron. If the alpha particle neared the massive, positively charged central body, the alpha particle (itself positively charged) was deflected. From the proportion of deflections, Rutherford could calculate the size of the nucleus. Nagaoka had no evidence of this type.

It is Rutherford, then, who rightly gets the credit for the advance. The central body is called the nucleus (plural nuclei) of the atom, from a Latin word meaning "little nut" in that it resembles a tiny nut inside the comparatively roomy atomic shell. Because, in biology, living cells also have central bodies called nuclei, that of the atom is sometimes specified as the atomic nucleus. For the purpose of this book, however, the qualifying word is not used.

Rutherford's picture of the nuclear atom proved entirely satisfactory, although many details have been filled in, as we shall see, in the three-fourths of a century since. For this and other work, Rutherford received a Nobel prize in 1908. (He received it in the category of chemistry, which displeased him, for he viewed himself as a physicist, of course.)

Positively Charged Particles

The nucleus has anywhere from 99.945 to 99.975 percent of the mass of the atom of which it is part. For this reason it became very important to study the nucleus. Indeed, you might almost say that the nucleus was the "real" atom. What had been thought of as the atom in the nineteenth century was mostly empty space; or at least, space filled with the very insubstantial particle/waves of the electrons. It was the nucleus that might have been the tiny, spherical, solid, and ultimate bit of matter that was first envisioned by Leucippus and Democritus.

Despite its mass, the nucleus is tiny in size, with a diameter only $\frac{1}{100,000}$ that of the atom. For that reason, it is considered as much a subatomic particle as the electron.

The nucleus must carry a positive electric charge; one of sufficient size to neutralize the charge of all of the electrons that are ordinarily to be found in a particular atom. Nevertheless, the history of such positively charged subatomic particles does not begin with Rutherford.

Goldstein, who had invented the name cathode ray, was interested in trying to find signs of any radiation traveling in the opposite direction. He could detect no such radiation emanating from an anode. In 1886, however, it occurred to him to devise a cathode that would itself allow radiation to travel in the other direction. This he tried to

The positively charged nucleus contains almost all of the mass of the atom, but it is only $1/100,000$ of its diameter.

achieve by making use of a cathode that was perforated and had little holes (or "channels") in it. When such a cathode was enclosed in the middle of an evacuated tube, and an electric current forced through it, cathode rays were formed. However, positively charged radiation, originating near the cathode, could pass through the channels, moving in the opposite direction.

This is precisely what Goldstein observed, and he called this new radiation *Kanalstrahlen*, which is German for "channel rays." However, this was incorrectly translated as "canal rays" in English.

In 1895, Perrin collected some of these canal rays on an object he placed in their path and showed that the object gained, in this way, a positive electric charge. In 1907, therefore, J. J. Thomson suggested they be called positive rays.

In 1898, Wien subjected these rays to magnetic and electric fields. He found that the particles of which positive rays were composed were much more massive than electrons. They were, indeed, as massive as atoms. In addition, the mass of the positive-ray particles depended on the traces of gas present in an evacuated tube. If it was hydrogen, the positive-ray particles had the mass of a hydrogen atom; if oxygen, they had the mass of an oxygen atom, and so on.

Once the basic theory of Rutherford's nuclear atom was accepted, it was immediately understood what the positive-ray particles were. The speeding electrons that made up the cathode rays collided with the stray atoms in the cathode-ray tube—hydrogen, oxygen, nitrogen, or whatever. The electrons were insufficiently massive to disturb the atomic nuclei and, in any case, struck them very rarely. They did, however, strike electrons and knock them out of the atom. The atoms, minus their electrons, would be nuclei carrying a positive electric charge, and would move off in the direction opposite to that taken by the cathode-ray particles.

As early as 1903, Rutherford had recognized that alpha particles were very similar to positive-ray particles in their properties. By 1908, he was quite certain that an alpha particle was just about equal in mass to the helium atom. It seemed to him there had to be some connection between alpha particles and helium because uranium minerals, which constantly produced alpha particles, also seemed, just as constantly, to contain small quantities of helium.

In 1909, Rutherford placed some radioactive material in a double-walled glass vessel. The inner glass wall was quite thin, but the outer glass wall was considerably thicker. In between the two walls was a vacuum.

The alpha particles ejected by the radioactive material could pass through the thin inner wall, but not through the thick outer wall. The alpha particles thus tended to be trapped in the space between the walls. After several days, the particles between the walls had accumulated to a volume at which they could be tested; and when this was done, helium was detected. It was clear, then, that alpha particles were helium nuclei. Other positive rays were nuclei of other types of atoms.

One of the ways in which positive-ray particles differed from electrons was that whereas all electrons had the same mass and the same electric charge, positive-ray particles had different masses and electric charges. Naturally, physicists wondered if they could somehow break up the positive-ray particles into smaller pieces and perhaps locate a very small positive particle no bigger than the electron.

Rutherford was among those who searched for such a tiny "positive electron," but didn't find it. The smallest positively charged particle he could find weighed as much as a hydrogen atom, and must be a hydrogen nucleus. In 1914, Rutherford decided that this particle must be the smallest positively charged particle that could exist. It has an electric charge exactly equal to that of the electron (although positive rather than negative), but has a mass, as we now know, 1836.11 times that of the electron.

Rutherford called this smallest positive-ray particle a proton, from the Greek word for "first," because when these particles are listed in order of increasing mass, the proton is first.

Atomic Numbers

Nuclei, which might be viewed as the essential cores of atoms, differ among themselves, as I have said, in two ways: in their mass and in the size of the positive charge they carry. This represents a significant advance over an earlier level of knowledge. Through the nineteenth century, nothing was known of the electric charges within the atom, and the only known difference among atoms was mass. Mass alone was not entirely satisfactory.

Earlier in the book, I mentioned that when the elements are arranged in the order of the mass of their atoms (atomic weight), a periodic table can be established. In this table, elements are so arranged that those with similar properties fall into the same row.

Such a table, based on atomic weight alone, has its faults. The size of the difference in mass varies as one goes up the scale. Sometimes the mass difference from one atom to the next is very small and sometimes quite large. In three cases, the atomic weight of a particular element is actually a bit *greater* than that of the element next higher in line.

In actual fact, if mass were all-important, the position of the two elements in these three cases ought to be reversed. They are not reversed, however, because, if they were, each of the elements involved would then be placed with a group that did not share its properties. Mendeleev, who first devised the periodic table, felt that keeping the elements with their own families was more important than strictly following the order of increasing atomic mass, and later chemists agreed.

Then, too, with only mass as a distinguishing characteristic, one could never be sure when an element might be discovered with an atomic weight in between two already known elements. As late as the 1890s, an entire family of elements, hitherto unknown, had been discovered and a new column had to be added to the periodic table. But it was possible that the confusing aspects of the periodic table might be done away with if the newly discovered second differentiating characteristic of atoms, the size of the positive charge on the nucleus, could be dealt with.

The possibility of doing so came by way of X rays. (There was no way of predicting, when X rays were first discovered, that they would be useful in connection with the periodic table. However, all knowledge is one. When a light brightens and illuminates a corner of a room, it adds to the general illumination of the entire room. Over and over again, scientific discoveries have provided answers to problems that had no apparent connection with the phenomena that gave rise to the discovery.)

The X rays first detected by Roentgen were produced when cathode rays impinged on the glass of a vacuum tube. The speeding electrons were suddenly slowed, and kinetic energy was lost. Such energy cannot be truly lost, but can only be converted into another form of energy; into electromagnetic radiation, in this case. The energy lost in a given moment was so great that unusually energetic photons were formed and the radiation was emitted in the form of X rays.

Once this was understood, it was quickly seen that if something denser than glass, and made up of more massive atoms, was put in the way of speeding electrons, they would be decelerated even more sharply. X rays would then be formed of still shorter wavelengths and higher energies. The obvious thing to use were various metal plates. These were placed at the opposite end of the tube from the cathode, where the speeding electrons would impinge upon them. Such metal plates were called anticathodes, where anti- is from a Greek word meaning "opposite." (Ordinarily, anodes are placed opposite the cathode, but to make room for an anticathode, the anodes are placed at the side of the tube.)

In 1911, the British physicist Charles Glover Barkla (1877–1944) noticed that when X rays were produced by an anticathode of a particular metal, they tended to penetrate substances just so far. Each metal produced X rays of penetrating power specific to the metal. Later on, when X rays were recognized as electromagnetic radiation, this was interpreted as meaning that each metal would produce X rays of a particular wavelength. Barkla called these the characteristic X rays of a particular metal.

Barkla also found that sometimes two types of X rays were produced by an anticathode of a particular metal, each with its own penetration, but with nothing much in between. He called the more penetrating beam K X rays, and the less penetrating one L X rays. Later on, still less penetrating beams were found to be produced in some cases, and the letters designating them continued through the alphabet, M X rays, N X rays, and so on. For his work, Barkla received a Nobel prize in 1917.

Barkla's work was carried on by one of Rutherford's students, Henry Gwyn-Jeffreys Moseley (1887–1915). In 1913, he studied the characteristic X rays very carefully, making use of the X-ray diffraction of crystals, which had just been discovered by the Braggs.

Moseley found that if he went up the list of elements in the periodic table, the wavelength of the X rays produced decreased regularly. The greater the atomic weight of the atoms in the anticathode, the shorter the wavelength of the X rays. Moreover, the change in wavelength was much more regular than the change in atomic weight.

Physicists were sure that the deceleration of electrons was brought about chiefly by the size of the positive charge on the atomic nucleus, which was an indication that the size of the charge as one went up the periodic table increased more regularly than did the mass of the atomic nucleus.

Moseley suggested, in fact, that the size of the charge increased by one with each step up the table. Thus, hydrogen, the first element, had as its nucleus the proton, which had a charge of $+1$. Helium, the second element, had a nucleus (the alpha particle) with a charge of $+2$. Lithium, the third element, had a nuclear charge of $+3$, and so on up to uranium, with the most massive atom then known, which had a nuclear charge of $+92$.

Moseley called the size of the nuclear charge the atomic number of the element, and this proved to be more fundamental than the atomic weight. Indeed, the atomic number solved many of the problems of the periodic table as Moseley's concepts were refined and extended by later physicists.

Thus, in those cases, in going up the periodic table, for which you have an element with a slightly higher atomic weight than the element following, this does not happen if atomic numbers are considered instead. An element that would seem out of place because it has a higher atomic weight than the one that follows turns out to have a lower atomic number. If all atoms are arranged by atomic number, every one, without exception, turns out to be with its proper family, and there need be no reversals. Then, too, when two neighboring elements have atomic numbers that differ by one, there can be no hitherto unknown element in between.

It soon became clear that all negative electric charges are exact multiples of the charge on an electron, while all positive electric charges are exact multiples of the charge on a proton. You can have nuclear charges of $+16$ and $+17$, but you can't have $+16.4$ or $+16.837$.

Where there is a missing element in the periodic table, the change in wavelength of the characteristic X rays, in

103

going from one element to the next over the gap, is twice as great as expected, and that is a sure sign of an element in between.

At the time Moseley worked out the concept of the atomic number, there were seven gaps in the table, each gap representing an as yet unknown element. By 1948, the gaps were all filled. Physicists were able to form atoms with atomic numbers higher than 92 by methods to be explained later. At the present time, all of the elements are known from atomic number 1 to atomic number 106. (Moseley would almost certainly have been awarded a Nobel prize for his work in this regard within a few years but, in 1915, he was killed in action in World War I at Gallipoli in Turkey.)

The atomic number tells us the size of the positive charge on the nucleus. Because the normal atom, as a whole, is electrically neutral, there must be one electron in the outer reaches of the atom for every positive charge on the nucleus. Thus, because hydrogen has a charge of $+1$ on the nucleus, the normal atom must possess one electron. Helium, with a nuclear charge of $+2$, must have two electrons in each atom; oxygen, with a nuclear charge of $+8$ must have eight electrons; uranium, with a nuclear charge of $+92$ must have ninety-two electrons, and so on. In short, the atomic number reflects not only the size of the nuclear charge, but the number of electrons in a normal atom.

It seems to make sense that chemical reactions take place when atoms, either independently, or as part of molecules, collide with each other. If so, the collisions are made basically between the electrons of one atom and the electrons of another. The nuclei of the two atoms are far off in the center of the atom, hidden behind the electrons, and are not at all likely to take part in chemical reactions or even to influence them in any crucial way.

This not only makes sense (things are not necessarily

so, just because they make sense), but seems to follow from such things as Arrhenius's theory of ionic dissociation. The formation of ions seems to be the result of the transfer of one or more electrons from one atom to another.

In the case of a molecule such as that of sugar, no ions seem to be formed. Instead, the atoms within the molecule simply cling together, perhaps because they share electrons and therefore cannot separate easily and remain intact atoms. However, there would seem to be some cases in which the transferring of electrons is the more stable situation, and others in which the sharing of electrons is—but why?

A hint comes from a group of six elements that are made up of atoms that do not tend to transfer or share electrons, but remain as single atoms at all times. The three lightest atoms in the group—of helium, neon, and argon— never transfer or share electrons at all, at least as far as chemists have ever observed. The three heaviest—of krypton, xenon, and radon—do share electrons under some extreme circumstances, but not very firmly.

These six elements are called the noble gases ("noble" because they are standoffish and do not tend to deal with the common herd). The reason for the "nobility" of these elements is best understood if we imagine that electrons are arranged about an atom in concentric shells, one outside the other. Naturally, as one moves outward from the nucleus, each successive electron shell is larger than the one before, and holds more electrons. Thus, the helium atom has two electrons, which seem to fill the innermost shell. This is not surprising, in that as the shell nearest the nucleus it should be the smallest of the shells, capable of holding the fewest electrons.

The American chemists Gilbert Newton Lewis (1875–1946) and Irving Langmuir (1881–1957), beginning in 1916, independently worked out notions of shells and electron transfer, or sharing, because these phenomena seemed to

105

explain chemical behavior so well. (Actually, the subject was greatly refined in later decades, but we'll get to that later.)

Individual shells are associated with the series of characteristic X rays first discovered by Barkla. The K series of X rays are the most penetrating and seem to originate from the electron shell nearest the nucleus. This first electron shell is therefore called the K shell.

Following this same reasoning, the shell just beyond the K shell is called the L shell because it seems to be the origin of the less-penetrating X rays of the L series. Beyond the L shell is the M shell, the N shell, and so on.

It might be that helium atoms are noble and will neither transfer nor share electrons (and therefore engage in no chemical reactions) because a filled shell is particularly stable. Either sharing or transferring an electron would lessen the stability of the situation, and stabilities are never lessened spontaneously. (It always takes energy to force something to destabilize, but stabilization takes place all by itself. These are properties associated with what is called the second law of thermodynamics.)

The next noble gas is neon, which has ten electrons in its atoms. The first two fill the K shell, and the next eight fill the L shell, which is larger and can hold more electrons. The electron pattern of neon is, therefore, 2, 8. With an L shell filled and stable, neon is a noble gas.

After neon is argon, with eighteen electrons in its atoms. These are arranged thus: two in the K shell, eight in the L shell, and eight in the M shell; or 2, 8, 8. The M shell, being larger than the L shell can hold more than eight electrons. Indeed, it can hold eighteen. However, eight electrons in the outermost shell (however many it can hold altogether) is a particularly stable configuration, making argon a noble gas.

After argon comes krypton, with thirty-six electrons,

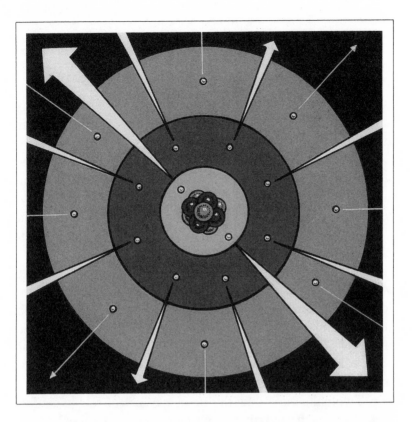

The innermost K shell, the middle L shell, and the outermost M shell of electrons in the noble gas argon, showing the relative strengths of X rays produced from each shell when struck by cathode rays.

arranged 2, 8, 18, 8; then xenon with fifty-four electrons, arranged 2, 8, 18, 18, 8; and finally, radon, with eighty-six electrons, arranged 2, 8, 18, 32, 18, 8.

Apparently, atoms interact with one another, when possible, so as to attain a noble gas configuration of electrons. Sodium, with eleven electrons, has them arranged 2, 8, 1. The eleventh electron is the only one in the M shell and is easily lost. When this happens, the sodium atom becomes a sodium ion, with one positive charge, because the charge of $+11$ on the nucleus is not completely neutralized with only ten electrons in the outer regions. (Notice that losing an electron does *not* change sodium into neon, which also has ten electrons in its atom. What counts, as far as the identity of the atom is concerned, is the nuclear charge, not the electron number.)

On the other hand, chlorine has seventeen electrons arranged 2, 8, 7. It needs one more electron to attain the noble gas configuration. It therefore has a tendency to gain one electron and become a negatively charged chloride ion with eighteen electrons in its atom, overbalancing the nuclear charge of $+17$.

For this reason, sodium and chlorine atoms easily react with each other. A transfer of one electron forms sodium ions and chloride ions, which cling to each other because positive and negative charges attract. When salt dissolves in water, the ions are less tightly bound and can slip past each other. In this way, such a solution will conduct an electric current.

Two chlorine atoms find another sort of stable configuration if each contributes an electron, so to speak, to a shared pool. Each atom has six electrons in its outermost shell that are entirely its own, and two electrons it shares with the other atom. Each outermost shell is full and stable provided the two atoms remain in contact so that they can maintain the two-electron sharing between them. The result is that the two chlorine atoms produce a two-atom chlorine molecule (Cl_2), which is more stable than two single chlorine atoms would be.

By thus dealing with electron arrangements in atoms, chemists found that they could understand why the periodic table is arranged as it is—based on chemical reactions that, in turn, depend on the electron arrangements in the outermost shell. Again, chemists found that they could use the electron arrangements to explain many chemical reactions that they had earlier simply accepted without knowing why.

It is, so far, sufficient to accept electrons as tiny, solid particles existing in geometric arrangements. However, such a view is insufficient in explaining the spectral lines that also distinguished each element from all others.

Spectral Lines

After Newton had demonstrated the existence of the light spectrum, it was studied closely by many scientists. If sunlight, for instance, is passed through a thin slit before being passed through a prism, each different wavelength casts an image of that slit in some characteristic color. The wavelengths line up very closely, seeming to form one smooth band of changing color (as in a rainbow). But what if some wavelengths happen to be missing for some reason? In that case, there will be places in the spectrum that produce no color image through the slit. That will produce a dark line across the spectrum.

In 1802, the British chemist William Hyde Wollaston (1766–1828) observed such dark lines, but he did not pursue the subject, and neither did anyone else for a while.

In 1814, however, the German optician Joseph von Fraunhofer (1787–1826) produced excellent prisms and other optical equipment, and was able to produce sharper spectra than anything produced before. At once he observed hundreds of dark lines in the spectrum. He carefully mapped their positions, and their prominence, and showed that the same lines fell in the same position whether their source was sunlight, moonlight, or light from the planets. (Of course, the light given off by the Moon and planets is reflected sunlight, so this is not, perhaps, surprising.)

From then on, the Fraunhofer lines, as they were frequently called, were studied carefully, but were regarded as little more than curiosities until an important breakthrough was made in 1859 by Kirchhoff.

Kirchhoff found that if particular elements were heated, they did not produce a continuous spectrum as the Sun did. Instead, they radiated light in separate wavelengths, so that the spectrum consisted of a number of bright lines, separated by stretches of darkness. If sunlight was sent through the relatively cool vapors of a particular element, the vapors would absorb just those wavelengths that they would emit when radiating. Moreover, each element emitted, when hot, or absorbed, when cool, its own characteristic wavelengths. In this way, the elements present in a particular mineral could be identified by the wavelengths emitted when the mineral was strongly heated. Elements hitherto unknown could be detected by the presence of wavelengths not given off by any known element. The elements present in the Sun and in other stars could be identified by the dark lines in their spectra.

All of this knowledge about spectral lines made them extremely important to chemists and to astronomers, but no one knew why different elements radiated or absorbed different wavelengths. One step forward in solving this puzzle was taken by the Swiss physicist Johann Jakob Balmer (1825–1898). He was particularly interested in the spectrum of glowing hydrogen, which seemed simpler than those of other elements (and why not, since hydrogen was the lightest and, presumably, the simplest of the elements).

The hydrogen spectrum consisted of a series of lines, spaced more and more closely with decreasing wavelength. In 1885, Balmer worked out a formula for the wavelengths of these lines. The formula contained a symbol that could be replaced by successive square numbers: 1, 4, 9, 16, and so on. As a result, the successive wavelengths of the lines in the hydrogen spectrum could be calculated. That still didn't explain why the lines were where they were, but at least it showed that there was a deep regularity in the lines that must somehow be reflected in the structure of the atom. There was no way of going further until more was known about the structure of the atom. Let's see how that worked out.

Once physicists accepted the nuclear atom, they had to consider what kept electrons in place. After all, if electrons are negatively charged and the nucleus is positively charged, and if opposite charges attract, why don't the electrons fall into the nucleus? The question might also be asked about the Earth—why doesn't it fall into the Sun, in that the two attract each other. In Earth's case, the answer is that it is in orbit. It *is* falling toward the Sun, but its additional motion at right angles to that fall keeps it forever in orbit.

There was a tendency to think, therefore, that the atom was a sort of miniature solar system, with the electrons whipping about the nucleus. There is a catch to that, however. From electromagnetic theory, it was known (and observed) that when an electrically charged object revolved in this fashion, it would give off electromagnetic radiation, losing energy in the process. As it lost energy, it would spiral inward and eventually fall into the nucleus.

In the same way, the Earth, in revolving about the Sun, gives off gravitational radiation, losing energy in the process, so that it is spiraling into the Sun. However, gravitation is so much weaker than electromagnetism that the amount of energy the Earth loses in this way is excessively small, allowing it to revolve about the Sun for billions of years without spiraling in appreciably closer.

An electron, however, subject to the much more intense electromagnetic field, loses so much energy in the form of radiation that its collapse into the nucleus, so it would seem, cannot be long delayed—yet this is not the

111

case. Atoms remain stable for indefinite periods, their electrons remaining in the outer portions.

This problem was tackled by the Danish physicist Niels Henrik David Bohr (1885–1962). He decided there was no use in saying that an electron radiated energy when it orbited an atom, when it clearly didn't. He insisted that as long as an electron remained in orbit, it *didn't* radiate energy.

Yet hydrogen, when heated, did radiate energy—and when cool, absorbed it. It emitted certain wavelengths that fit Balmer's equation, and absorbed those same wavelengths. To explain this, Bohr supposed, in 1913, that the electron in the hydrogen atom could take on any of a number of different orbits at different distances from the nucleus. Whenever it was in a particular orbit, whatever its size, it didn't either gain or lose energy. When the electron changed orbits, however, it either absorbed energy, if it moved farther from the nucleus, or emitted energy, if it moved closer to the nucleus.

But why should an electron be in a particular orbit and then, absorbing energy, suddenly shoot outward into the next larger orbit—never, by any chance, being in an orbit halfway between the two. Bohr saw that this had to have some connection with the quantum theory. If the atom could only handle quanta of a certain size, it could only absorb light of a certain wavelength, and that would automatically send it outward to the next orbit.

Bohr worked out a series of calculations that showed how one could map a series of permitted orbits that would result in the absorption or emission of quanta of fixed size (and hence, radiation of fixed wavelength) that would perfectly account for the particular wavelengths of the lines in the hydrogen spectrum.

What Bohr did was to show that one could not work out the structure of the atom solely according to classical

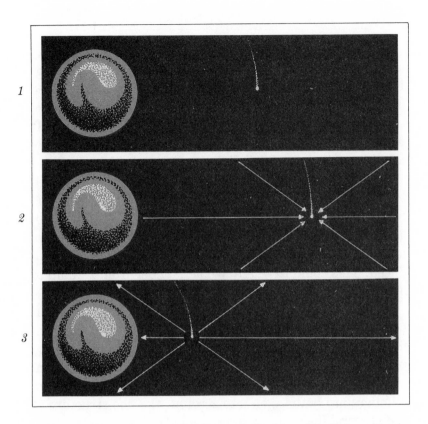

1. *Whenever an electron stays in a particular orbit, it doesn't gain or lose energy.*
2. *When it changes to a higher orbit, it absorbs energy.*
3. *It emits energy when it moves closer to the nucleus.*

physics; one had to make use of quantum theory. For this he received a Nobel prize in 1922.

Bohr had to use whole numbers for one of the terms of his formula, each number representing a set of different spectral lines. The need for whole numbers arose out of the fact that a whole number of quanta were involved. You couldn't have fractions of a quantum. The number that was inserted in the formula was called a quantum number for that reason.

Although Bohr's formula gave the figures for the wavelengths of the spectral lines, it didn't explain everything. If the spectral lines were studied with increasingly refined instruments, it turned out that each line had a "fine structure"—a number of very closely spaced thinner lines. It was as though each of Bohr's orbits consisted of a family of orbits with very small differences among themselves.

In 1916, the German physicist Arnold Johannes Wilhelm Sommerfeld (1868–1951) pointed out that Bohr's orbits were all circular. The orbits might be elliptical, too, and to different degrees. In order to take these new orbits into account, a second quantum number had to be introduced. This could have any whole number from zero up to one less than Bohr's quantum number.

If Bohr's quantum number (or principal quantum number) is 1, the lowest it can be, Sommerfeld's quantum number (or orbital quantum number) can only be 0. If the principal quantum number is 2, then the orbital quantum number could be either 0 or 1, and so on. If the two quantum numbers are both taken into account, the fine structure of the spectral lines can be expressed.

Complications continued, however. If atoms are in a magnetic field, lines that seemed absolutely single split into finer parts. Both Bohr and Sommerfeld pictured the orbits (whether circular or elliptical) as all being in a single plane, so that the nucleus and all possible orbits formed a system that was as flat as a sheet of paper. It was possible, though, for orbits to be tipped, so that all of the orbits taken together might be distributed symmetrically through three-dimensional space, and the atom given a spherical outline. This makes sense, because atoms do, in many ways, act as though they were tiny spheres.

In order to take the three-dimensional system of orbits into account, a third quantum number had to be added—the magnetic quantum number. This could have any positive whole value from 0 to whatever the principal quantum number might be, and any negative integer value of the same sort. If the principal quantum number were 3, the magnetic quantum number might be -3, -2, -1, 0, $+1$, $+2$, or $+3$.

With three dimensions taken care of, it might seem there was nothing further that needed to be done. Nevertheless, there were still certain features of the spectral lines that were puzzling, and one more quantum number was added by the Austro-Swiss physicist Wolfgang Pauli (1900–1958). This was thought to represent the spin of the electron on its axis. This spin could be in one direction or the other, clockwise or counterclockwise. In order to make the calculations fit the observed facts about the spectral lines, this spin quantum number had to be either $+\frac{1}{2}$ or $-\frac{1}{2}$.

Pauli went on to show that within an atom no two electrons could exist with all four quantum numbers identical. This is called the exclusion principle because once an electron has its four quantum numbers, any other electron is excluded from the particular orbit represented by those numbers. For this, Pauli received a Nobel prize in 1945. (Sometimes, a scientist must wait twenty years, or, in rare cases, even as long as fifty years, for a Nobel prize. It takes time, now and then, to see that a discovery is truly significant. If prizes were awarded immediately for something that *seemed* important, many would be given for discoveries that would turn out to be trivial or even wrong.)

A detailed system of mathematics that uses the four quantum numbers and the exclusion principle to describe how electrons are distributed in an atom was worked out by the Italian physicist Enrico Fermi (1901–1954) in 1926, and by the British physicist Paul Adrien Maurice Dirac (1902–1984) in 1927. This system is called Fermi-Dirac statistics, which applies to any particle that has a spin of $+\frac{1}{2}$

or $-\frac{1}{2}$. Such particles are lumped together as fermions, for Fermi. The electron is an example of a fermion. So is the proton.

There are particles that have spins of 0, 1, or 2. (The photon has a spin of 1, for instance, and the graviton has a spin of 2.) The exclusion principle does not hold for such particles, whose manner of distribution was worked out by the Indian physicist Satyendra Nath Bose (1894–1974) in 1924. Einstein praised Bose's work and, in 1925, added to it. This system is called Bose-Einstein statistics, and any particle with an integral spin, or one of zero, is called a boson, for Bose.

Bohr's electron orbits, although an enormous advance, were nevertheless not entirely satisfactory. There was still the image of electrons as particles in motion, racing about their orbits. If this is so, there is still no clear explanation of why the electron isn't emitting radiation and spiraling into the nucleus. It is all very well to say that while an electron is in orbit it doesn't radiate; but why doesn't it? It is also a compelling argument to say that it can only emit quanta of a certain size; but why? Something is missing.

The German physicist Werner Karl Heisenberg (1901–1976) thought there would always be trouble if one attempted to picture the structure of the atom in terms of ordinary everyday life. What we are used to—planets circling the Sun, or billiard balls striking each other—involves masses so large compared with the atom that the tiny quanta out of which energy is built are too small to have any noticeable effect on such objects. All of our mental images, therefore, are of a nonquantum world. In dealing with atoms, electrons, and radiation, however, we are dealing with a world in which quantum effects are noticeable, so our images fail. (The quantum theory is, in a way, a system of saying that the Universe is grainy, not totally smooth. It is like a newspaper photograph that looks smooth

Taken to the highest level of magnification, the substance of the Universe is grainy, not smooth.

because the little dots of black and white of which it is made up are too small for us to see. If we magnify the picture sufficiently, our ordinary-world images fail. All we see are the dots, which no longer form a discernible picture.)

It seemed to Heisenberg that one should use only the figures taken from the spectra and manipulate them in some way that would enable physicists to work out numerical answers useful in connection with atomic behavior, rather than to attempt to interpret the behavior in terms of orbits, ellipses, tilts, spin, and all of the rest of it. In 1925, Heisenberg worked out what was called matrix mechanics, for

117

In dealing with the world in which quantum effects are noticeable, all our conventional images fail. If we consider the electron to be a wave rather than a particle, it might appear to form a wavy hoop about the nucleus.

the purpose, because it made use of a mathematical device called a matrix.

In that same year, however, Davisson had proved the existence of electron waves, and it occurred to the Austrian physicist Erwin Schrödinger (1887–1961) that these waves could explain the nature of electron orbits.

If we consider the electron to be a wave, rather than a particle, we can picture an orbit about the nucleus as having to consist of whole-number wavelengths. Then, if

118

we imagine ourselves tracing the wave around the nucleus, it doubles back on itself to form a "path" that looks like a wavy hoop about the nucleus. The smallest orbit would be that consisting of a single wavelength up and down. The electron couldn't spiral into the proton because it couldn't take up an orbit with a length less than a single wave. All of the other orbits would be at distances and with shapes such that a whole number of waves would fit around the orbit. That is why orbits can only be at certain distances, in certain ellipses, at certain tilts, with certain spins, and so on.

Schrödinger worked out a mathematical treatment that could solve problems by taking electron waves into account, and announced it in 1926. His system is called wave mechanics. Dirac also contributed to the working out of the treatment, and he and Schrödinger shared a Nobel prize in 1933 for this work.

It turned out eventually that wave mechanics and matrix mechanics are equivalent; they deliver the same results. The mathematical system is, therefore, simply called quantum mechanics. With improvements and refinements added, the system has proven completely satisfactory in dealing with electrons, and with subatomic phenomena generally.

In 1939, the American chemist Linus Carl Pauling (b. 1901) applied the principles of quantum mechanics to the manner in which atoms transferred and shared electrons. This replaced the older particle system of Langmuir and Lewis; it was subtler and explained many things the older system could not. For this, Pauling received a Nobel prize in 1954.

Earlier, in 1927, Heisenberg had demonstrated that it was impossible, even in principle, to work out certain types of measurement with perfect accuracy because of the graininess of the Universe implied by quantum mechanics. For example, suppose you tried to determine the exact position of a particle, as well as its exact momentum (which is its mass multiplied by its velocity). Any device you might use to determine position would change the particle's velocity and, therefore, its momentum. Any device you might use to determine momentum would change the particle's position. The best you could do would be to get the combined position and momentum with a tiny fuzziness; with a small, unavoidable inaccuracy. The uncertainty of the position multiplied by the uncertainty of the momentum, if both are taken at the absolute minimum that can be obtained, comes to an amount closely related to a fundamental constant of quantum theory.

Heisenberg's uncertainty principle also says that one can't determine time and energy content simultaneously and exactly. For this work, Heisenberg received a Nobel prize in 1932. The uncertainty principle is a very important discovery, explaining a great deal that would otherwise be mysterious in subatomic physics. Nevertheless, many scientists shied away from it because it made it seem as though there was a certain element of randomness in the Universe that could never be wiped away. Einstein, for instance, never accepted the uncertainty principle, and always thought that quantum mechanics was an incomplete theory because of it.

Still, not wholly accepting the uncertainty principle doesn't remove it. Moreover, the uncertainty principle seems to describe the Universe as it *is*, and there's no point in fighting it.

Bohr's picture of electron orbits seemed to describe an electron particle whose position and motion could, and should, be known at every moment. Schrödinger's use of waves, which proves to work much better, does *not* do this. An electron wave goes up and down, and somewhere in it is the electron in its particle aspect. We can't tell exactly where the particle is, however. In a way, it's everywhere along the wave. The height of the wave tells us the probability that it would happen to be there at any given moment, but it doesn't *have* to be there. In this way, quantum mechanics features probability and uncertainty—and that, indeed, seems to be the way the Universe is.

Because quantum theory deals with things so far removed from what we are used to in ordinary life, scientists speak of "quantum weirdness." There are aspects about it that seem so paradoxical that scientists have simply not managed to agree on what it all means. Perhaps someday, new discoveries, new concepts, new thoughts will clarify what seems now to be hopelessly mysterious.

The game of science never stops, you see, for new problems arise whenever old problems are solved; but who would want it any other way? To solve everything would stop the game, and nothing else that life could offer would, in my opinion, make up for the intellectual loss.

5

ISOTOPES

Nuclear Energy

Working out the details concerning the electron distribution in atoms was, in a way, a rather simple problem. All electrons are alike, and no differences among them, whether they are in one type of atom or another, or exist independently, have been discovered to this day. Atoms differ among themselves in the number of electrons they possess, but not in the type of electrons.

What about atomic nuclei, though? They differ from atom to atom in both mass and electric charge. Are they single particles of many different types, one for each different element, or do they have internal structure? Are they built out of different numbers of simpler particles, and are these simpler particles the same in all nuclei of all elements? Moreover, are these questions we can practically hope to answer? After all, nuclei are tiny objects hidden at the very center of the atom, sometimes behind layers and layers of electrons. How would one reach and study them?

The first hints concerning nuclear structure came with the discovery of radioactivity, fifteen years before the existence of the atomic nucleus was demonstrated. One question that naturally arose at once in connection with radioactivity was where all of that energy might be coming from. Uranium seemed to ceaselessly emit alpha rays, which were streams of helium nuclei, and beta rays, which were streams of electrons. All of these particles traveled at very high speeds, alpha particles at about a tenth the speed of light, and beta particles at about nine-tenths the speed of light. It takes considerable energy to make them move that quickly from a standing start. (After all, the uranium atoms aren't moving to begin with.) Then, too, there were gamma rays, which were far more energetic than even X rays.

Uranium's radiation isn't just a brief spurt of energy. A sample of uranium metal will continue to radiate indefinitely at an apparently constant rate, and this was a serious problem. By the law of conservation of energy, it would seem that energy could not be created out of nothing, and yet energy *seemed* to be created out of nothing in connection with radioactivity.

Of course, it might have been that the law of conservation of energy was wrong, or was limited to only certain conditions. Scientists, however, found the law so useful in all aspects of science that they hated to scrap it. There was the definite feeling that a search must be made to explain radioactivity *without* giving up the law of conservation of energy; that the law must be given up only as an absolutely last resort. (This is an example of intelligent conservatism

123

among scientists. A theory or a law that has proved itself over and over should not lightly be discarded. It should be, and would be, discarded if there were no alternative, but one must be sure that there were indeed no alternatives.)

The situation grew quickly worse in the years immediately following the discovery of radioactivity. Marie Curie and her husband, Pierre Curie (1859–1906), started with pitchblende, a uranium-containing rock, from which they hoped to obtain samples of pure uranium for study. To their astonishment, they found that pitchblende was more radioactive than it would be even if it were made of pure uranium. Possibly, it contained elements still more radioactive than uranium. There was no sign of such elements in ordinary analysis, so they must exist in very small quantities, and, if so, must be *very* radioactive.

In 1898—after long, tedious, and painstaking work that began with tons of pitchblende and ended with tiny pinches of radioactive powder—the Curies isolated two elements: polonium (named for Marie Curie's native Poland) and radium (named for its radioactivity). Each was far more radioactive than uranium.

If one wondered at uranium's energy emission, how much more one must wonder at radium, which gave off energy at nearly three million times the rate. In 1901, Pierre Curie measured the energy given off by radium and found that 1 gram of radium gave off energy at a rate amounting to 140 calories per hour. That wasn't much in itself, but it continued for hour after hour indefinitely. Where did all that energy come from?

Some scientists wondered if radioactive atoms might not absorb energy from the surrounding environment and convert it into the energy of radiation. This hypothesis, however, would break the second law of thermodynamics, and scientists were as reluctant to do this as to break the first law (conservation of energy).

124

In 1903, Rutherford suggested that all atoms possessed large volumes of energy within their structure. Ordinarily, this energy was never tapped, so that people remained unaware of its existence. Radioactivity was, however, a spontaneous outpouring of a little of this energy. This was a daring suggestion, but it caught the imagination of the public and people began to speak of atomic energy, as a newer and far more concentrated form of energy than had ever before been known. (The English writer H.G. Wells even wrote of "atomic bombs" in his science fiction stories forty years before such a thing existed in reality.)

And yet Rutherford's suggestion might have seemed to be a case of pulling a rabbit out of a hat. Just *saying* that the atom contained energy explained nothing. But then, in 1905, Einstein showed, convincingly, that mass was a very concentrated form of energy. If radioactive substances were to turn even a small fraction of their mass into energy, then all of the energy liberated in radioactivity could easily be accounted for.

Once the nuclear atom was discovered, it was clear that because almost all the mass of an atom was concentrated in the nucleus, the necessary loss of mass must take place there. It was within the nucleus, then, that the energy source of radioactivity lay, and, eventually, people began to speak of nuclear energy rather than atomic energy.

Nuclear Varieties

If the energy of radioactivity arises out of the loss of mass of the atomic nucleus, what happens to that nucleus as a result? The beginnings of an answer came even before it was understood that the atomic nucleus was the source of energy, or even that there was an atomic nucleus at all. (It

125

often happens that scientific observations provide the beginnings of an answer to a question even before the establishment of a comprehensive theory that provides order and reason to a section of science. Such early observations are difficult to understand, and advance knowledge slowly without the theory. Once the theory is established, however, the earlier observations quickly fall into place. Advance is then rapid until it is slowed by the absence of some deeper and broader understanding of another aspect of science.)

In 1900, Crookes, working with uranium metal, decided to purify it as much as he could, and subjected it to chemical proceedings that separated out apparent impurities. He found, to his astonishment, that the purified uranium was hardly radioactive at all, while the impurities were markedly radioactive. He suggested that it was not uranium that was radioactive, but something else that existed in uranium as an impurity.

Becquerel, however, having discovered the radioactivity of uranium, was not ready to let go of it that easily. (Scientists often treat their discoveries as their babies, and defend them vigorously against any attempt to wipe them out. This is a very human reaction, even if, sometimes, in hindsight, wrong. In this case, though, Becquerel was proved to be right.) Becquerel showed that uranium, when purified in Crookes's fashion, did indeed show little radioactivity, but if this purified uranium were allowed to stand, it would regain its radioactivity after some time.

In 1902, Rutherford and a co-worker, the British chemist Frederick Soddy (1877–1956), showed that this was also true of thorium. If the metal were purified, most of its radioactivity was lost, but was then regained on standing. Rutherford and Soddy suggested, therefore, that when an atom of uranium gave off radioactive radiations, its nature was changed and it became an atom of another element that was more radioactive. This new element, being radioactive, also changed. Uranium was not very radioactive in itself, but its daughter elements were. When uranium was purified so that the daughter elements and their radioactivity were removed, uranium seemed far less radioactive than it had been, but it slowly formed additional quantities of daughter elements and its radioactivity returned and became as before. Atoms, it seemed, in undergoing radioactivity, experienced something that could be described as radioactive disintegration.

As it turned out, this was a correct view. Both uranium and thorium broke down into other elements, which, in turn, broke down until finally an element was reached that was not radioactive. In this way, one had a radioactive series. Scientists began to search for these intermediate elements in disintegrating uranium and thorium. Polonium and radium, earlier detected by the Curies, were two of them—along with others. It was also discovered that both uranium and thorium, after undergoing many changes, become nonradioactive lead.

The notion of radioactive disintegration came as a shock to scientists. After all, from the time of Leucippus and Democritus on it had been assumed that atoms were unchangeable—but that had only been an assumption. To be sure, atoms are unchangeable as far as chemical changes are concerned, but radioactivity is not a chemical change. Chemical changes involve only the outermost electrons of an atom. The atom might gain an electric charge, or form a bond with another atom as a result, but its essential identity, which depends upon its nucleus, remains intact. Radioactivity, however, *does* involve the nucleus. It is a nuclear change and, if the nucleus undergoes a change, it is quite likely that, in the process, one type of atom changes into another.

(A change in point of view, such as this, does not mean that all chemical textbooks need to be torn up and thrown

127

out as though all of the information they contain is now worthless. The new point of view merely broadens and extends knowledge, and supplies explanations that are fuller and more useful. Thus, twentieth-century textbooks must take into account the existence of nuclear change, but they can, if they wish, still discuss *chemical* change, as before, and treat atoms as changeless—as, indeed, they are, where chemical change is concerned.)

The search for the radioactive intermediates between uranium and lead, and between thorium and lead, was successful—too successful, actually. Far too many were found.

The atomic number of uranium is 92, and that of thorium 90. The atomic number of lead is 82, and another known element, bismuth, has an atomic number of 83. The as yet undiscovered elements lying at the end of the periodic table were elements numbered 84, 85, 86, 87, 88, 89, and 91. That's seven altogether, minus the newly discovered elements polonium (84) and radium (88), which left five. There were no other elements beyond the remaining five to be discovered between uranium and lead. None! That was quite certain after Moseley's work in 1914.

And yet by Moseley's time, more than thirty intermediates had been discovered. Each one of these was distinctly different, at least as far as its radioactive properties were concerned. Some would expel an alpha particle and some a beta particle. Some would expel a gamma ray with or without an alpha particle or a beta particle. Even when each of two intermediates emitted an alpha particle, say, one would do so with greater energy than the other, and more rapidly.

Soddy tackled this problem. Already in 1912 and 1913, before the concept of the atomic number had been worked out, he had found that certain intermediates had the same chemical properties, and, if mixed, could not be separated by ordinary chemical procedures. They were the same elements, and that meant (as was later understood) that their

128

electron configurations were the same, as was the positive charge on their nuclei. Because they differed in their radioactive properties, however, something about the nucleus, other than the charge, was *not* the same.

The periodic table was based on the chemical properties of the elements. It followed that if two different atoms were identical in chemical properties and different only in radioactive properties, they were the same element (chemically) and must both fit into the same place in the periodic table.

Soddy announced his findings in 1913, naming these different types of atoms—which were of the same element and belonging in the same place in the periodic table—*isotopes*, from Greek words meaning "same place." He received a Nobel prize for this work in 1921.

This was yet another blow to long-held views about atoms. Leucippus, Democritus, and Dalton had all assumed that all of the atoms of a particular element were identical. There hadn't seemed to be any observations that pointed to the contrary—until now. Scientists working with radioactive intermediates found as many as five or six varieties of an element, each with distinct radioactive properties.

Once the concept of atomic number became clear in 1914, it was possible to see the details of how one type of atom changed into another. Thus, the uranium atom has a nucleus with an atomic weight of 238, and an atomic number of 92. We call it U-238. However, in its radioactive transformation it gives off an alpha particle, which has an atomic weight of 4 and an atomic number of 2. The atomic weight and atomic number of the alpha particle must be subtracted from that of the uranium nucleus. What is left, then, is a nucleus with an atomic weight of 234, and an atomic number of +90. (An alpha particle, when emitted, always reduces the atomic weight of the emitting nucleus by 4 and the atomic number by 2.)

When this disintegration of the uranium nucleus was

discovered by Crookes, he called the product uranium X, which was just a way of saying he hadn't the slightest idea what it could be. But now, with the cold figures of the change, it could be seen that the new atom was thorium, all atoms of which, after all, have an atomic number of 90.

Ordinary, well-known thorium has an atomic weight of 232 and is, therefore, Th-232. The product of uranium disintegration has an atomic weight of 234 and is Th-234. Here we have an example of two isotopes. Both possess an atomic number of 90 and have a nuclear charge, therefore, of +90. There is, however, a difference in mass. Th-234 is two units more massive than Th-232.

Does this really make a difference? Chemically, it doesn't. Both Th-232 and Th-234, having a nuclear charge of +90, have 90 electrons in their atoms, arranged in the same way in each case, so that all chemical properties are the same. However, from the radioactive standpoint, it does make a difference. Thorium-232, the ordinary thorium we find in minerals, gives off alpha particles, while thorium-234, the product of uranium disintegration, gives off beta particles. Moreover, the atoms of thorium-234 disintegrate some 200 billion times as rapidly as the atoms of thorium-232. That's quite a difference.

There are other isotopes of thorium that turn up as part of one radioactive series or another. They include Th-227, Th-228, Th-229, Th-230, and Th-231. They all break down in different ways and at different rates, and all do so much more quickly than Th-232. But let's go back to thorium-234, as it gives off a beta particle. Does it change as a result?

A beta particle is an electron. It has a charge of −1, so it can be considered to have an atomic number of −1. Its mass is 1/1,837 that of a hydrogen atom, or about 0.00054. This is so small a figure that we won't go far wrong if we consider it as just about 0. This means that if a nucleus emits a beta particle, one must subtract 0 from the nucleus's atomic weight—leaving it unchanged. We must also subtract -1 from the nucleus's atomic number. Subtracting -1 is equivalent to adding $+1$, so the atomic number goes *up* by 1. Therefore, the Th-234 nucleus with an atomic number of 90 and an atomic weight of 234, emitting a beta particle, is changed to a nucleus having an atomic number of 91 and an atomic weight of 234. The element of atomic number 91 is protactinium, which was first isolated and identified in 1917 by the German chemist Otto Hahn (1879–1968) and his co-worker, the Austrian chemist Lise Meitner (1878–1968). What we have, then, is Th-234 changed to Pa-234.

The emission of a gamma ray by an atomic nucleus does not change the nucleus. The gamma ray has an atomic number of 0 because it has no charge, and an atomic weight of 0 because it has no mass. A nucleus, in emitting a gamma ray, merely loses energy.

Once scientists knew how each of the radioactive radiations changed an atomic nucleus, they were able to work out the precise identity of all intermediates in a radioactive series.

The concept of isotopes left the periodic table intact. Each place contained only one type of atom as far as atomic number was concerned. That isotopes differed in atomic weight did not matter where chemical properties were concerned. What it meant in connection with nuclear structure and properties we will come to later.

Half-Lives

The various intermediates in a radioactive series break down quite quickly. If a given quantity of one of these intermediates is observed, it will be seen that the number of breakdowns declines with time. The reason is clear. As the atoms break down, fewer and fewer of the original variety are left to break down, and the fewer further breakdowns there are to be observed.

The manner in which the rate of breakdown declines is precisely what is to be expected of something that chemists were familiar with in the case of many chemical reactions. It is what is called a first-order reaction. This means that each radioactive atom of a particular variety has a certain chance of breaking down, and that this chance doesn't change with time. It might have one chance in two of breaking down on a particular day, but if a hundred days pass without its having broken down, it still has only one chance in two of breaking down on the 101st day. (This is analogous to the situation with respect to tossing a coin. You have one chance in two of tossing heads. Still, if you toss the coin a hundred times and get tails each time, the chance of tossing a head on the 101st time is still only one in two—assuming, of course, that it is an honest coin. It is often erroneously believed that the more times one tosses a tail, the greater the chance of a head on the next occasion.)

You can't tell when an individual atom will break down, but, if you are dealing with a great many atoms, you can calculate how many will break down in the course of a day, or a minute. You won't know *which* atoms will break down, but you will know the number. This is similar to the way in which statisticians can predict how many motorists are likely to die on a holiday weekend, even though they can't possibly tell which particular motorists will die.

This means that you can calculate how long it will take for half of all atoms present to break down. It turns out, in the case of first-order reactions, that it always takes the same time for half of any quantity to break down. Thus, if you start with 120 grams of a given isotope, and if it takes

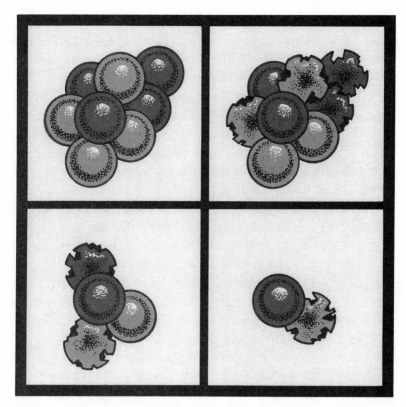

Half-life is the average time required for one-half of the atoms of a radioactive material to undergo radioactive decay.

a year for half of it to break down, it will take another year for half of the remaining half to break down. Put another way, starting with 120 grams, you will have 60 grams at the end of one year, 30 at the end of two, 15 at the end of three, 7.5 at the end of four, and so on. Theoretically, you will never get to zero, but, eventually you will have a single atom and, after some unpredictable period of time, it, too, will break down and your radioactive isotope will be gone.

In many cases, scientists can count the number of actual alpha particles or beta particles given off per unit of time. From the way in which that number falls off, they can calculate the amount of time it would take half of the isotope to be gone. (Scientists have worked out various ways of detecting individual alpha and beta particles, but I will discuss devices in this book only when I must. What I am concentrating on is ideas and concepts.)

Thus, the protactinium isotope obtained from uranium breakdown, Pa-234, will lose half its atoms in about 70 seconds. This is its half-life, a term introduced by Ernest Rutherford in 1904.

Naturally, protactinium-234, if it existed all by itself, would be gone in not much time, even if there were enormous quantities of it to begin with. If the entire Earth consisted of nothing but protactinium-234, and if it could be imagined that the atoms would break down quietly, this vast amount would be gone in about three hours. (Actually, so much energy would be produced that the Earth would explode like an enormous bomb.)

And yet protactinium-234 does exist in the Earth's soil, and can be isolated in very small quantities. Why isn't it *all* gone? The answer is that any such atoms that existed when the Earth was formed were gone a few minutes later; however, more are constantly being formed from uranium.

Other isotopes have longer half-lives. Radium-226 (the isotope isolated from pitchblende by the Curies), which emits alpha particles, has a half-life that is quite long, so that over short periods of time the decline in breakdown rate is too small to notice. If one waits long enough, however, the decline can be measured, and it turns out that the half-life is 1,620 years. But even this is not long enough for radium to endure for the lifetime of Earth. Radium exists only because it is constantly being formed from uranium.

Because uranium has a very slow rate of breakdown, radium forms very slowly. Radium, however, breaks down as it forms, but quite slowly at first because there is so little of it. As more and more accumulates, however, it breaks down faster and faster (characteristic of a first-order reaction), eventually breaking down as rapidly as it forms and reaching radioactive equilibrium.

It turns out that in any mineral that contains uranium, there will also be radium, with the amount of radium much smaller than uranium because of radium's shorter half-life. This is so even if uranium doesn't produce radium directly, but only through several other intermediate stages.

As it happens, the concentration of uranium in uranium ore is 2,780,000 times greater than that of radium, so the half-life of uranium-238 is 2,780,000 times longer than that of radium-226. This means that the half-life of uranium-238 is about 4.51 billion years.

This is why there is still primordial uranium on Earth. The Earth was first formed about 4.6 billion years ago, and included a certain amount of uranium in its makeup to begin with. In all that vast length of time, only about half of that primordial uranium has broken down. The other half is still here. It will take another 4.51 billion years for half of what is now left to break down. It is because uranium has been around all this time that the intermediates of its breakdown are in existence as well, although in much smaller quantities, of course.

Thorium-232 has an even longer half-life than uranium—13.9 billion years. Only about ⅕ of the original supply of thorium on Earth has yet had a chance to break down.

There is the uranium isotope, uranium-235, discovered in 1935 by the Canadian-American physicist Arthur Jeffrey Dempster (1886–1950). It is not nearly as long-lived as uranium-238 or thorium-232. The half-life of uranium-235 is only 710 million years. That is still long enough, however, to allow a little over ⅟₇₀ of the original quantity present at Earth's beginning to exist today.

Stable Nuclear Varieties

Soddy's discovery of isotopes involved only radioactive atoms, yet his discovery immediately cast suspicion on non-radioactive atoms. As early as 1905, the American chemist Bertram Borden Boltwood (1870–1927), noting that uranium minerals always seemed to contain lead, wondered if lead might not be the final product of radioactive disintegration. As investigation proceeded, this proved to be so. This meant that lead, although a nonradioactive element, was intimately involved with radioactivity.

The one way a radioactive atom ordinarily changes its atomic weight is by emitting an alpha particle. A beta particle affects the atomic weight insignificantly, and a gamma ray does so not at all. Every time an alpha particle is emitted, the atomic weight decreases by four. This means that if the original radioactive atom had an atomic weight evenly divisible by four, all of the intermediate products, without exception, would also have to have an atomic weight evenly divisible by four—as would the final lead atom. Thus, thorium-232, the only long-lived thorium isotope, has an atomic weight divisible by 4 (232 = 58 × 4). As its breakdown proceeds, it loses a total of 6 alpha particles with a total atomic weight of 24, leaving an atomic weight of 208 for what remains of the nucleus. These six alpha particles also cause thorium-232 to lose a total of 12 positive charges; however, 4 beta particles are emitted, which restores 4 positive charges. The net loss in positive charges is, therefore, 8.

Thorium has an atomic number of 90. Losing 8 positive charges produces an atom with an atomic number of 82, which is that of lead. Consider the atomic weight loss of 24, and you see that the final product of the disintegration of thorium-232 is lead-208, which is not radioactive but stable, and of which there is a reasonable quantity on Earth— and always has been, and always will be.

That's fine, so far, but consider uranium-238. Its atomic weight, when divided by 4, leaves a remainder of 2 (238 = 59 × 4 + 2). If it loses atomic weight by emitting alpha particles, all of its intermediate products as well as its final product will have atomic weights that when divided by 4 will leave a remainder of 2. An atom of uranium-238, in its disintegration, loses 8 alpha particles and 6 beta particles, which makes it lead-206.

Finally, uranium-235 has an atomic weight that, when divided by 4, leaves a remainder of 3 (235 = 58 × 4 + 3), as do all of its intermediates and its final product. Each uranium-235 atom gives off 7 alpha particles and 4 beta particles, ending up as lead-207. (There is a fourth series in which all of the atomic weights, when divided by 4, leave a remainder of 1. We will have occasion to mention it later.)

We are left, then, with three different lead isotopes: lead-206, lead-207, and lead-208. Each is stable, and each possesses the usual properties of lead. Which of these, then, if any, exists in nature independent of radioactivity?

Suppose we consider the atomic weight of lead. As found in nature, in rocks that have no suspicion of radioactivity about them, lead has an atomic weight of 207.19. In that the various stable isotopes are always present in a fixed proportion, could this number simply be an average atomic weight? (Because all of the various geological processes depend on the chemical properties of the various minerals, they cannot separate isotopes as they separate the various elements according to their chemical properties, but leave the isotopes thoroughly mixed in the same proportions at all times.)

Let's test the proposition. Suppose you had a rock rich in uranium. In addition to the original supply of lead, if any, you would have a slow, constant addition of lead-206

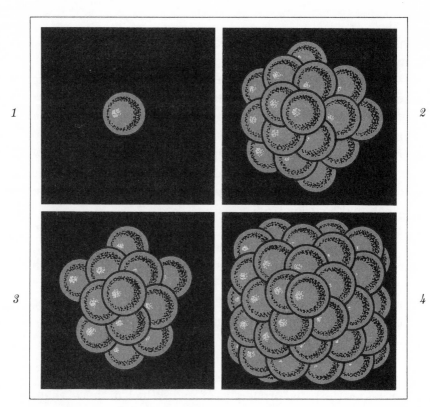

In 1815, William Prout suggested that all atoms are combinations of hydrogen atoms.
(1) Hydrogen 1. (2) Carbon 12. (3) Oxygen 16. (4) Sulfur 32.

and lead-207, making the atomic weight of lead in the rock measurably less than it would be in nonradioactive rock. A rock rich in thorium would be undergoing slow, constant addition of lead-208, making the atomic weight of its lead content higher than it would be in nonradioactive rock.

In 1914, Richards measured the molecular weight of lead taken from various radioactive ores. He found that thorium minerals gave values as high as 207.9 for lead, while uranium minerals gave values as low as 206.01.

In the same year, then, that atomic numbers had replaced atomic weights as fundamental to the periodic table, it suddenly began to appear that atomic weights were not fundamental at all. They might merely be averages of isotopic weights (mass number), which might themselves be much more significant.

But, of course, lead isotopes were prepared through radioactive breakdown. Perhaps that is a special case. What about elements that have nothing whatever to do with radioactivity? There were hints on this point even before the existence of lead isotopes were virtually proved by Richards's findings.

Suppose we consider positive rays, which are streams of positively charged atoms possessing less than their normal number of electrons. (Sometimes positive rays contain no electrons, consisting simply of bare nuclei.) If these positive-ray particles are placed in electromagnetic fields, their paths curve away from the straight lines they normally move in. The extent of curvature depends on both the electric charge of the particles and their mass. If we're dealing with an element whose atoms all have the same number of electrons removed, then all of the particles that make up the ray have the same positive electric charge. Therefore, if we witness any deviation in the curvature of the ray path, it must be because of differences in particle mass—that is, in their atomic weights.

Suppose the gas in a tube were neon, all of whose atoms carried the same positive charge. If all of the atoms also had the same atomic weight (as had always been taken for granted since the atomic theory had been established), they would all curve along the same path. If photographic film were placed in the way of the speeding particles, they would all strike the film at the same place, forming a small, fogged spot.

Thomson tried the experiment in 1912 and found that the neon ions did fog the photographic plate nearly in the expected position; however, very close to it there was a second, considerably less prominent spot of fogging. The position at which the second fogging took place was about that expected of atoms with an atomic weight of 22. No atomic weight of this size was expected, but Thomson suggested that if, out of every ten neon atoms, nine had a mass number of 20 and one of 22, the weighted average of the two would come out to 20.2, which was close to the measured atomic weight of neon as it occurs on Earth. In other words, neon, which had nothing to do with radioactive processes, might be made up of two isotopes: neon-20 and neon-22. This possibility suddenly opened up a new view of nuclear structure.

In 1815, in the infancy of the atomic theory, the British chemist William Prout (1785–1850) had suggested (anonymously because the idea was too far-out for him to dare attach his name to it) that all atoms are combinations of hydrogen atoms. Atomic weights were just being determined, and they seemed to be whole numbers. That is, hydrogen was 1, carbon was 12, oxygen was 16, sulfur was 32, and so on. Prout suggested that the carbon atom was made up of 12 hydrogen atoms in close association, the oxygen atom of 16 hydrogen atoms, the sulfur atom of 32 hydrogen atoms, and so on.

This suggestion was called Prout's hypothesis when the authorship was revealed. It didn't stand up, however, for as atomic weights were determined more and more accurately, it turned out that they were by no means all integers, or even close to them. Chlorine, for instance, was 35.456; copper was 63.54; iron was 55.85; magnesium was 24.31; mercury was 200.59; and so on.

Prout's hypothesis, if true, would have made the atomic theory much more elegant; that is, simpler and neater. However, observations had forced its abandonment for a century. Now it was suddenly back in the forefront of thinking.

If a positively charged ionic stream of an element is placed in an electromagnetic field, the less massive isotopes will be more easily deflected from their usual straight-line path. In a mass spectrometer, this effect will produce closely spaced dark lines on a photographic plate. The position of the line allows the mass number of the isotope to be calculated, and the darkness of the line indicates the relative amount of that isotope.

What if all atomic weights that were not integers were simply the averages of the mass numbers of various isotopes—mass numbers that *were* integers. If so, atomic weights might be useful in chemical calculations, but it would be the isotopic mass numbers that would be useful in considering nuclear structure.

In 1919, the British chemist Francis William Aston (1877–1945), who had been a student of J. J. Thomson's, devised what he called a mass spectrograph. This caused charged ions with the same charge and mass to be concentrated in a fine line on a photographic plate. In this way, the presence of isotopes could be seen as closely spaced dark lines. The position of a line allowed the mass number of an isotope to be calculated, and the darkness of the line the relative amount of that isotope. The results would be much more precise than those obtained by Thomson's groundbreaking but crude instrument.

Using such a mass spectrograph, the lines for neon-20 and neon-22 were clearly detected—as was, eventually, a very faint line for neon-21. We now know that out of every 1,000 neon atoms, about 909 are neon-20, 88 are neon-22, and 3 are neon-21. All three isotopes are stable, with the weighted average of their mass numbers giving neon, as found in nature, an atomic weight of 20.18. For his work with the mass spectrograph, Aston received a Nobel prize in 1922.

Other elements were, of course, tested, and a majority of them found to consist of several isotopes. Chlorine, for instance, is made up of two isotopes: chlorine-35 and chlorine-37. Of every 1,000 chlorine atoms, 755 are chlorine-35 and 245 are chlorine-37, making the weighted average of their mass numbers just about the atomic weight of chlorine as it exists in nature. (The weighted average doesn't come out *exactly* to the measured atomic weight because the mass numbers, as we shall see, aren't *quite* integers, either.)

Sometimes one isotope is present in an overwhelming majority. Out of every 1,000 atoms of carbon, for instance, 989 are carbon-12 and 11 are carbon-13. Out of every 1,000 nitrogen atoms, 996 are nitrogen-14 and 4 are nitrogen-15. Out of every 10,000 hydrogen atoms, 9,999 are hydrogen-1 and 1 is hydrogen-2. Out of every million helium atoms, all but one are helium-4, the odd one being helium-3. In all of these cases, the atomic weight is close to an integer.

In 1919, the American chemist William Francis

Giauque (1895–1982) discovered that out of every 10,000 oxygen atoms, 9,976 are oxygen-16, 20 are oxygen-18, and 4 are oxygen-17. What made the finding significant was that oxygen had been used as the standard for atomic weight since Berzelius's time, its atomic weight being set exactly equal to 16.0000. However, that was now seen to be only an average that could vary slightly from sample to sample. In 1961, therefore, physicists and chemists officially agreed to tie the standard to a mass number and not to an atomic weight. The mass number of carbon-12 was set to 12.0000 exactly, which shifted the old atomic weights only very slightly. For instance, the atomic weight of oxygen is no longer 16.0000, but 15.9994.

Some atoms come in only one variety in nature. Thus, all fluorine atoms in nature have a mass number of 19, all sodium atoms 23, all aluminum atoms 27, all cobalt atoms 59, all gold atoms 197, and so on. For these cases, many physicists believed the word *isotope* to be inappropriate. Isotope implies that at least two atomic varieties are packed into the same place in the periodic table. To say that an element has one isotope is like saying that a parent has one twin offspring. In 1947, therefore, the American chemist Truman Paul Kohman (b. 1916) suggested that the term *nuclide* be used instead. It is a perfectly good term, but isotope has become too firmly established to be displaced.

There are eighty-one elements, each of which has at least one stable isotope. Of these, the most complicated is bismuth, which has an atomic number of 83. All of its atoms have a mass number of 209. The most massive stable atom, then, is that of bismuth-209.

There is no stable atom with an atomic number greater than 83 or an atomic weight greater than 209. More massive atoms exist on Earth only because uranium-238, uranium-232, and thorium-232, although radioactive, are very long-lived.

The total number of stable isotopes distributed among

the eighty-one elements is 272, which is enough to supply three or four per element, if they were distributed evenly. They are not, of course. The elements with even atomic number generally have a greater than average supply. Tin, with an atomic number of 50, holds the record with ten stable isotopes, with mass numbers of 112, 114, 115, 116, 117, 118, 119, 120, 122, and 124.

Elements with odd atomic number generally have either one or two stable isotopes. There are nineteen elements (all but one with an odd atomic number) that are made up of a single stable isotope. The one exception, with an even atomic number, is beryllium (atomic number 4), which has a single stable isotope, beryllium-9.

You might wonder why it is that there are only eighty-one elements with stable isotopes, if elements numbers 82 (lead) and 83 (bismuth) have them. Clearly, there must be two elements without them in the list of elements between 1 and 83, and this is so. Elements 43 and 61 (both odd atomic numbers) do not have any stable or nearly stable isotopes. They were sought for diligently in the 1920s and, occasionally, they were reported to have been isolated from one ore or another, but all such reports turned out to be mistaken. Neither element was actually isolated until scientists learned to form nuclei in the laboratory that don't exist in measurable quantities on Earth itself. (We'll get to such matters later on.)

Another peculiarity is that potassium (atomic number 19) is the only element of odd atomic number to have more than two isotopes occurring in nature. It has three, with mass numbers 39, 40, and 41. Of these three, however, potassium-40 makes up only about 1 out of every 10,000 potassium atoms.

As early as 1912, Otto Hahn had noted that potassium seemed to be weakly radioactive, and, eventually, this was narrowed down to potassium-40. Potassium-40 is long-lived, with a half-life of 1.3 billion years. This is a longer half-life than that of uranium-235. Less than a tenth of the quantity present in the Earth when it first formed still exists today. Potassium, however, is so common an element in Earth's rocks that even though only 1 out of 10,000 potassium atoms is potassium-40, there is more potassium-40 in rocks than uranium-238 and uranium-235 combined.

If this is so, why wasn't radioactivity discovered in potassium sooner than it was in uranium? The answer is that, first, uranium emits energetic alpha particles, and potassium-40 only rather feeble beta particles. Second, uranium breaks down into a long series of intermediates, each of which is more strongly radioactive than uranium itself. Potassium-40, on the other hand, breaks down directly into a stable isotope, argon-40.

Potassium-40 is not the only nearly stable isotope among those we have listed as stable. There are about a dozen others, all of which have half-lives much longer than potassium-40, or even than thorium-232. They are so long-lived that their radioactivity can barely be detected. Vanadium-50, for instance, has a half-life of about 600 trillion years, which is some 130,000 times as long as that of uranium-238; neodymium-154 has a half-life of about 5,000 trillion years, and so on. None of these nearly stable isotopes with atomic numbers less than that of thorium (90) gives rise to a series of breakdown products. All but one give up a single beta particle and become a stable isotope. Samarium-147 is the exception, giving up a single alpha particle to become the stable neodymium-143.

The fact that the mass numbers of isotopes are all very close to integers made it very tempting to think that nuclei are (as Prout had suggested) made up of smaller particles, and that the various particles to be found within nuclei might be very few. The chance of simplifying nature in this way was very enticing, and in the 1920s physicists worked hard at puzzling out the structure of the atomic nucleus.

145

6

NEUTRONS

Protons and Electrons

The desire for simplicity was not the only force driving physicists to look into the realm of nuclear structure. From actual observations of radioactive materials, it seemed clear that at least some nuclei *must* have a structure; that is, be a collection of still simpler particles. Thus, some radioactive nuclei gave off beta particles (electrons) and others gave off alpha particles (helium nuclei). The simplest explanation for these emissions was that nuclei contained within themselves simpler nuclei and electrons, which were, for some reason, occasionally released.

If we are satisfied that some nuclei are made up of smaller nuclei plus electrons, then it is an easy jump to

146

speculate that all nuclei might possess this structure. For simplicity's sake, we might also suppose that nuclei, when composed of simpler nuclei, should be composed of the simplest nuclei possible.

The smallest nucleus known was that of hydrogen-1, a nucleus with a mass number of 1 and an electric charge of $+1$. Rutherford had called the hydrogen-1 nucleus a proton, and there was the general feeling in the 1920s that the proton was the smallest and simplest particle capable of carrying a positive charge. There arose the theory, then, that atomic nuclei might be made up of protons and electrons crushed together into a tiny volume.

The alpha particle, which emerges from some radioactive atoms, has a mass number of 4, so it could be made up of 4 protons, each with a mass number of 1. However, the alpha particle also has an electric charge of $+2$, and 4 protons have a total charge of $+4$. It would seem, then, that in addition to the 4 protons in the alpha particle, there must be 2 electrons, canceling two of the positive charges while adding nothing significant to the mass. A 4-proton/2-electron alpha particle would then have, as observed, a mass of 4 and a charge of $+2$.

This sort of thing could be worked out for other nuclei, too. It could be used to explain isotopes. For instance, the oxygen-16 nucleus has a mass of 16 and a charge of $+8$, so it should consist of 16 protons and 8 electrons. The oxygen-17 nucleus could be viewed as having an additional proton-electron pair, which increases the mass by 1, without changing the charge. A total of 17 protons and 9 electrons yields a mass of 17 and a charge of $+8$. Again, the oxygen-18 nucleus can be viewed as having still another proton-neutron pair, so that it would be made up of 18 protons and 10 electrons, with a mass of 18 and a charge of $+8$.

For a while, physicists rode high on this proton-electron theory of nucleus structure, particularly because it

Protons and neutrons are both present in the nucleus.

reduced the universe to extraordinary simplicity. All material objects, so the theory stated, in the universe are made up of about 100 types of atoms, and every atom, in this view, is made up of equal numbers of two types of subatomic particles: protons and electrons. All protons are located in the nucleus, while some electrons are located in the nucleus and others outside the nucleus.

Furthermore, it seemed that the entire universe was held together by two fields. The nucleus was held together by the electromagnetic attraction between protons and electrons; the atom, as a whole, was held together by the electromagnetic attraction between nuclei and electrons. Various atoms were combined to form molecules, or crystals, or solid objects as large as planets by the transfer of electrons from one atom to another, or by the sharing of electrons. Was there anything not held together by electromagnetic fields? Yes, of course.

The molecules of gases are scattered far apart and are subjected to only very feeble electromagnetic forces. But if this were the only force exerted on them, the molecules would dissipate and distribute themselves throughout the vastness of space. Gases, however, are held to a large body through the influence of something else: gravitational attraction. That is why our atmosphere clings to the Earth.

The gravitational field, however, is so weak that it takes a very large body to hold gases. For instance, low-boiling liquids on Earth would tend to evaporate and their molecules flee into space if gravitation were not strong enough. It is because of Earth's gravitational pull that we have an ocean, while the Moon is not large enough to have free water on its surface.

Bodies separated by considerable distances in space are also held together by gravitational fields: satellites to planets, planets to stars, stars to one another to form galaxies, and galaxies to one another to form clusters. The Universe as a whole is, indeed, held together by gravitational attraction.

Add to this the fact that the electromagnetic field is associated with a radiation of photons, and the gravitational field with a radiation of gravitons, and it would seem that the entire Universe consists of but four types of particles: protons, electrons, photons, and gravitons. Protons have a mass number of 1, a charge of $+1$, and a spin of $+\frac{1}{2}$ or $-\frac{1}{2}$. Electrons have a mass number of 0.00055, a charge of -1, and a spin of $+\frac{1}{2}$ or $-\frac{1}{2}$. Photons have a mass number of 0, a charge of 0, and a spin of $+1$ or -1. Gravitons have a mass number of 0, a charge of 0, and a spin of $+2$ or -2.

How simple that is! It is even simpler than the Greek

notion of four elements that applied to Earth and a fifth element assigned to the heavenly bodies. As a matter of fact, the Universe was never again to seem quite as simple as it seemed for a few years in the 1920s.

Indeed, there was a mighty attempt to make it simpler still. Why should there be two fields: electromagnetic and gravitational? Might these not be two aspects of the same phenomenon? Might not a single set of equations describe both?

To be sure, the electromagnetic field and the gravitational field seem utterly different. The electromagnetic field involves only electrically charged particles, while the gravitational field involves all particles with mass, charged or not. The electromagnetic field involves both attraction and repulsion, while the gravitational field involves only attraction. The electromagnetic field is trillions of trillions of trillions of times as intense as the gravitational field for a given pair of particles that respond to both. Thus, in considering a proton-electron pair, we need only take into account the electromagnetic attraction between them; the gravitational attraction is, in comparison, insignificant.

Nevertheless, such differences need not be a bar to unification. Magnetism, electricity, and light seemed, at first, to be three widely different phenomena, and yet Maxwell found a set of equations that held for all three and showed them to be different aspects of the same phenomenon.

None other than Einstein spent the final decades of his life trying to complete the work of Maxwell by finding still more fundamental equations that would include the gravitational field as well, in what was called a unified field theory. He failed, but, as we shall see, that did not end the attempts.

The proton-electron system of nuclear structure itself did not hold up. It contained a fatal flaw.

Protons and Neutrons

The nucleus has a spin, just as electrons, protons, photons, and gravitons do. The amount of spin can be determined by a close study of the fine lines of the spectrum produced by given nuclei, and by other methods.

If the nucleus is made up of constituent particles such as protons and electrons, then it stands to reason that the total nuclear spin is the sum of the spins of the constituent particles. This is because the spins represent angular momentum, and for a long time physicists have found that there is a law of conservation of angular momentum. In other words, you can't create spin out of nothing, or destroy it. It can only be transferred from one body to another.

This holds for all ordinary bodies as far as the law can be tested. An ordinary spinning object (such as a coin you set spinning with your hand) might seem to get its spin from nowhere. The spin, however, comes from the motion of your hand, and when you twist the coin, your hand, the rest of your body, and whatever you are attached to—a chair, the ground, the planet Earth—gets a reverse twist. (Angular momentum can be in either of two directions, plus and minus, and the two can cancel each other out. In addition, one can be formed out of nothing if the other is simultaneously formed. It is the *net* angular momentum—what you get when all the pluses and minuses are added together—that is conserved.)

The amount of angular momentum depends not only on the speed of turning, but on the mass of the turning object. When you twist your hand to set a coin spinning, the Earth is so much more massive than the coin that its reverse twist takes place at a speed far too small to measure by any conceivable method. And when the coin slows its spin through friction with a surface and finally stops spin-

151

ning, the Earth's reverse spin, incomprehensibly small, also slows and stops.

In the case of spinning particles, the spin is potentially eternal if the particles are left to themselves. Both protons and electrons have spins that can be represented by half-integers, $+\frac{1}{2}$ and $-\frac{1}{2}$. (The total spin is the same for both particles despite the difference in mass. The electrons just spin faster to make up for their lesser mass. Of course, the spin can be in either direction.)

If the spins of an even number of protons and electrons within a nucleus are added, the total spin has to be zero or an integer. Two spins, for instance, can be $+\frac{1}{2}$ and $+\frac{1}{2}$, or $+\frac{1}{2}$ and $-\frac{1}{2}$, or $-\frac{1}{2}$ and $+\frac{1}{2}$, or $-\frac{1}{2}$ and $-\frac{1}{2}$. The sums are respectively $+1$, 0, 0, and -1. If you imagine four half-integer spins, or six, or eight, or any even number and add them with whatever combination of pluses and minuses you please, you will always come out with zero, a positive integer, or a negative integer.

If you have an odd number of particles, each with a half-integer spin, you end up with a half-integer spin no matter how you shift the pluses and minuses. If there are three particles, for instance, you can have $+\frac{1}{2}$ and $+\frac{1}{2}$ and $+\frac{1}{2}$, for a total of $+1\frac{1}{2}$; or $+\frac{1}{2}$ and $+\frac{1}{2}$ and $-\frac{1}{2}$, for a total of $+\frac{1}{2}$. However you shift the pluses and minuses for three particles, or five, or seven, or any odd number, you always end either with $+\frac{1}{2}$, $-\frac{1}{2}$, $+$ some integer and $\frac{1}{2}$, or $-$ some integer and $\frac{1}{2}$.

This brings us to the nitrogen-14 nucleus, which was shown by spectroscopic studies to have a spin of either $+1$ or -1. The nitrogen-14 nucleus has a mass number of 14 and an electric charge of $+7$. By the proton-electron scheme of nuclear structure, its nucleus must be made up of 14 protons and 7 electrons, or 21 particles altogether. However, 21 particles, being an odd number, must add up to a total spin of a half-integer. They cannot add up to either $+1$ or -1.

This bothered physicists very much. They didn't want to give up the proton-electron picture of the nucleus because it was so simple and explained so much, but they didn't want to give up the law of angular conservation, either.

As early as 1920, some physicists, notably Ernest Rutherford, wondered if a proton-electron combination could be viewed as a single particle. It would have the mass of a proton (or a tiny bit more, thanks to the electron), with an electric charge of zero.

Of course, you couldn't consider such a particle to be just a proton and electron fused together because each would contribute a spin of $+\frac{1}{2}$ or $-\frac{1}{2}$ to the fusion, whereby the fused particle would have a spin of either 0, $+1$, or -1. The spin of the nitrogen nucleus would then still be a half-integer altogether, whether you count the protons and electrons separately, or in combinations with each other.

You must instead think of a particle that has a mass of 1, like a proton, a charge of 0, and a spin of $+\frac{1}{2}$ or $-\frac{1}{2}$. Only so could the requirement of the nitrogen nucleus be met. In 1921, the American chemist William Draper Harkins (1871–1951) applied the name *neutron* to such a particle, because it was electrically neutral.

Throughout the 1920s, this possibility remained in the minds of physicists, but because the hypothetical neutron was never actually detected, it was difficult to take the position seriously. The proton-electron scheme continued in use, therefore, even though it didn't fit all of the facts of reality. (Scientists don't generally abandon a notion that seems useful until they are sure they have a better one to put in its place. Replacing something useful with nothing, or with something very vague, is not a good idea in scientific procedure.)

In 1930, the German physicist Walter W.G.F. Bothe (1891–1957) reported that when he bombarded the light element beryllium with alpha particles, he got a radiation of some sort. It was very penetrating and it didn't seem to carry an electric charge. The only thing he could think of that had these characteristics were gamma rays, so he suspected that that was what he had.

In 1932, the French physicist Frédéric Joliot-Curie (1900–1958) and his wife, Irène Joliot-Curie (1897–1956), the daughter of Pierre and Marie Curie, found that Bothe's radiation, when it struck paraffin, brought about the ejection of protons from that substance. Gamma rays were not known to do that, but the Joliot-Curies could not think of any other explanation.

The British physicist James Chadwick (1891–1974), however, repeated the work of Bothe and the Joliot-Curies, in 1932, and reasoned that in order for the radiation to eject a massive particle such as a proton, it had to consist of massive particles itself. Because it certainly didn't carry an electric charge, he decided that here was the massive, neutral particle physicists were searching for—the neutron. That was indeed what it was, and in 1935 Chadwick received a Nobel prize for his discovery.

Once the neutron was discovered, Heisenberg immediately suggested that the atomic nucleus consisted of a closely packed mass of protons and neutrons. The nitrogen nucleus, for instance, was made up of 7 protons and 7 neutrons, each with a mass number of 1. The total mass number was, therefore, 14, and because only the protons had a $+1$ charge while the neutrons had a charge of 0, the total charge was $+7$, as it was supposed to be. Moreover, there were now 14 particles altogether—an even number—thus the total spin of the nucleus could be either $+1$ or -1, as measured.

It turned out that the proton-neutron structure explained, without exception, the nuclear spin of all atomic nuclei, while at the same time explaining everything the proton-electron structure had explained (with one excep-

154

tion that was later filled in, as I will eventually explain). Indeed, in the better than half a century since the discovery of the neutron, nothing has been found that would in the least shake the proton-neutron structure of the nucleus, although there have been refinements of the idea, which we will come to later.

Consider, for instance, how neatly the new notion explains the existence of isotopes. All of the atoms of a given element have the same number of protons in the nucleus and, therefore, the same nuclear charge. The number of neutrons, however, might differ.

Thus, the nitrogen-14 nucleus is made up of 7 protons and 7 neutrons, but one out of 3,000 nitrogen nuclei contains 7 protons and 8 neutrons and is, therefore, nitrogen-15. Although the most common oxygen nucleus contains 8 protons and 8 neutrons in its nucleus, which makes it oxygen-16, a few have 8 protons and 9 neutrons, or even 8 protons and 10 neutrons (oxygen-17 and oxygen-18, respectively).

Even hydrogen, with a nucleus made up of a single proton and nothing else (hydrogen-1), is not immune. In 1931, the American chemist Harold Clayton Urey (1893–1981) showed that 1 out of 7,000 hydrogen atoms was hydrogen-2, and he received a Nobel prize in 1934 for this work. The hydrogen-2 nucleus consists of 1 proton and 1 neutron, which is why it is often called deuterium, from the Greek word for "second."

Similarly, uranium-238 has a nucleus made up of 92 protons and 146 neutrons, while uranium-235 has one made up of 92 protons and 143 neutrons. There isn't an isotope of any type that doesn't fit in perfectly with the proton-neutron nuclear structure.

Protons and neutrons are both present in the nucleus (they are sometimes lumped together as nucleons), both are of almost equal mass, and both can, under the proper conditions, be ejected from the nucleus. Yet the proton was

155

recognized as a particle in 1914, while the neutron had to wait an additional eighteen years for its discovery. Why did it take so long to discover the neutron? The reason is that electric charge is the most easily recognized portion of a particle, and while the proton carries an electric charge, the neutron does not.

One of the earliest ways of recognizing the existence of subatomic particles was by means of the gold-leaf electroscope. This device consists of two thin and very light sheets of gold leaf attached to a rod and enclosed in a box designed to protect the assemblage from disturbing air currents. If an electrically charged object is touched to the rod, the charge enters the gold leaf. Because both sheets of gold leaf receive the same charge, they repel each other and move apart in an inverted V.

Left to itself, the electroscope would remain with its leaves apart. Any stream of charged particles entering the electroscope, however, will knock electrons from molecules of air. This produces negatively charged electrons and positively charged ions (a stream of charged particles is an example of ionizing radiation). One or the other of these charged particles will neutralize the charge on one of the gold leaves, causing them to come slowly together. A stream of neutrons, however, is not an example of an ionizing radiation because, being uncharged, they neither attract nor repel the electrons out of atoms and molecules. Neutrons, therefore, cannot be detected by the electroscope.

The German physicist Hans Wilhelm Geiger (1882–1945) invented a device, in 1913, that consisted of a cylinder containing a gas under a high electric potential, but one that was not quite high enough to force a spark of electricity through the gas. Any bit of ionizing radiation entering the cylinder would form an ion, which would be pulled through the cylinder by the electric potential, creating more ions.

156

Even a single subatomic particle produced a discharge that would make a clicking sound. The Geiger counter became famous as a way of counting subatomic particles.

Even earlier, in 1911, the British physicist Charles Thomson Rees Wilson (1869–1959) invented a cloud chamber. He allowed dust-free moist air to expand in a cylinder. As it expanded, it cooled, and some of the moisture would condense into tiny droplets, provided there were dust particles present as centers about which the droplet could form. Without dust particles, water remains in vapor form. If a subatomic particle entered the cloud chamber, it would form ions all along its path, and these would act as water condensation centers. About each ion a tiny water droplet would form. In this way, not only the particle, but its pathway as well, could be detected. If the cloud chamber were placed in an electric or magnetic field, the speeding charged particle would curve in response—its path visible. Wilson received a Nobel prize in 1927 for this device.

In 1952, the American physicist Donald Arthur Glaser (b. 1926) invented a similar device. Instead of a gas out of which liquid droplets were ready to form, Glaser used a liquid raised to a temperature at which vapor bubbles were about to form. Those bubbles formed along the pathway of an entering subatomic particle. For this "bubble chamber," Glaser received a Nobel prize in 1960.

All of these devices, and many others like them, respond to the formation of ions by ionizing radiation; that is, by electrically charged particles. None of them will work for neutrons, which enter and leave the devices silently, so to speak.

The presence of neutrons can be detected only indirectly. If a neutron formed inside a detecting device travels a distance and then collides with some other particle that *can* be detected—provided the neutron alters the pathway of the other particle or forms new, detectable particles—

157

there will be a gap between the pathways that mark the formation of the neutron at one end and the collision of the neutron with something else at the other. That gap has to be filled with something, and from the nature of the two sets of pathways that can be seen, it is logical to deduce the presence of a neutron in between.

Physicists who work with particle-detecting devices learn to photograph complex pathways marked out in drops of water, bubbles of gas, lines of sparks, and so on, and interpret all of the details as easily as we might read this book.

It is because neutrons leave no marks in these devices that caused their discovery to be delayed for so many years. Once found, however, they proved to be of enormous importance, as we will see if we now move back a little in time.

Nuclear Reactions

All of the innumerable interactions of atoms and molecules that involve transfers and sharings of electrons are called chemical reactions. Until 1896, all of the interactions scientists knew about, either in living tissue or in inanimate nature, were chemical reactions, although their nature was not really understood, of course, until the structure of atoms came to be known.

In this respect, radioactivity is different. The changes involved in radioactivity involve the ejection of portions of the nucleus, or of changes in the nature of particles within the nucleus. Such events are called nuclear reactions, which, in general, involve much greater intensities of energy change than do chemical reactions.

Radioactivity is a spontaneous nuclear reaction. If

158

there weren't just a few spontaneous nuclear reactions, those that take place without any initiation or interference by human beings, it is possible we might never have discovered the existence of such things.

It is, after all, much more difficult for human beings to initiate or control nuclear reactions than it is to do the same with chemical reactions. To produce, prevent, or modify a chemical reaction, chemists need only mix chemicals, or heat them, cool them, put them under pressure, blow air through them, or carry out other easily managed procedures. After all, it is only the outer electrons that are involved, and they are so exposed that we can easily fiddle with them.

Nuclear reactions take place in the tiny nuclei at the very center of atoms—nuclei that are shielded by numbers of electrons. All of the procedures used to bring about chemical change do not reach or affect the atomic nuclei. Thus, when radioactivity was first discovered, chemists were astonished to find that the rate of breakdown was not altered by temperature change. Whether a radioactive substance was heated to melting, or placed in liquid air, the rate of radioactivity continued as before, unchanged. Subjecting a radioactive substance to chemical change did not alter the rate of radioactive breakdown, either.

Was there any way, then, of interfering with the nucleus at all? If such a method existed, it would have to involve reaching through the electronic shield and, so to speak, touching the nucleus itself. It was precisely in this way that Rutherford had discovered the existence of the nucleus. He had bombarded atoms with energetic alpha particles, which were massive enough to brush electrons aside, and small enough to bounce away from a nucleus as it approached.

In 1919, Rutherford placed a bit of radioactive material at one end of a closed cylinder, with an inner coating of

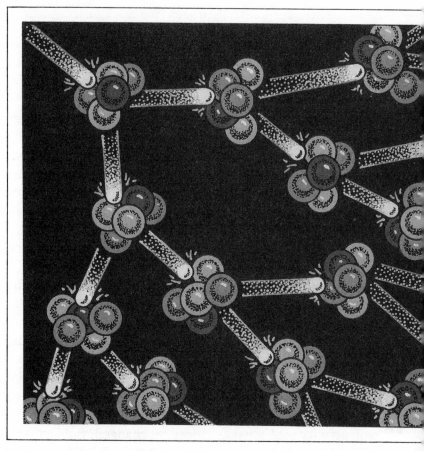

Neutrons collide with nuclei, causing more and more neutrons to be emitted in a nuclear chain reaction.

zinc sulfide at the other end. The radioactive material emitted alpha particles. Whenever an alpha particle struck and was stopped by the zinc sulfide, the alpha particle lost its kinetic energy, which was converted into a tiny flash of light that could just be seen if the room were kept very dark and the eyes were allowed to adapt to the darkness. By counting the scintillations of light, Rutherford and his co-workers could actually count the individual particle strikes. Such a device is called a scintillation counter.

160

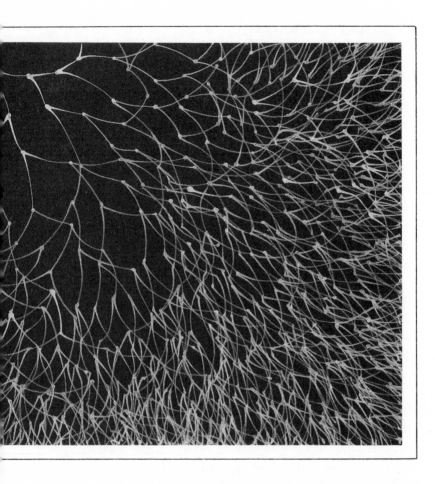

If alpha particles are passed through a vacuum, the scintillations are many and bright. If, however, some hydrogen is allowed in the cylinder, particularly bright scintillations appear. This would seem to be because the alpha particles occasionally strike the proton nuclei of the hydrogen, and the protons, being lighter than the alpha particles, can be knocked forward at a greater speed. Speed counts for more than mass where kinetic energy is concerned, so the very speedy protons produce very bright scintillations.

161

If oxygen or carbon dioxide are allowed into the cylinder, the scintillations grow dimmer and fewer. The comparatively massive nuclei of the oxygen and carbon atoms (respectively, four and three times the mass of the alpha particles) tend to slow down the alpha particles, sometimes to a point at which they pick up electrons and become ordinary helium atoms. The massive carbon and oxygen nuclei are knocked forward slowly, and what scintillations there are, are dim.

If, however, nitrogen is placed in the cylinder, the bright scintillations one observes with hydrogen appear. Rutherford supposed that the particles in the nitrogen nucleus were less firmly bound together than those in either the carbon or oxygen nucleus. The alpha particle could bang against the carbon or oxygen nuclei without shaking them apart, but when it collided with a nitrogen nucleus, it would knock a proton out of the nucleus, causing the usual proton scintillation.

This was only speculation at first, but, in 1925, the British physicist Patrick Maynard Stuart Blackett (1897–1974) put the Wilson cloud chamber to wholesale use for the first time in order to check up on Rutherford's experiment. He bombarded nitrogen in a cloud chamber with alpha particles, and took 20,000 photographs, catching a total of more than 400,000 alpha particle tracks. Of these, just eight involved a collision between an alpha particle and a nitrogen molecule.

By studying the tracks going into the collision and coming out of it, Blackett showed that Rutherford was right, that a proton had been knocked out of the nitrogen nucleus. The alpha particle, with a charge of $+2$, had entered the nucleus, and a proton, with a charge of $+1$, had left the nucleus. That meant the nucleus had a net increase of $+1$ in its charge. Instead of being $+7$ (nitrogen), it had become $+8$ (oxygen). Furthermore, the alpha particle had entered

162

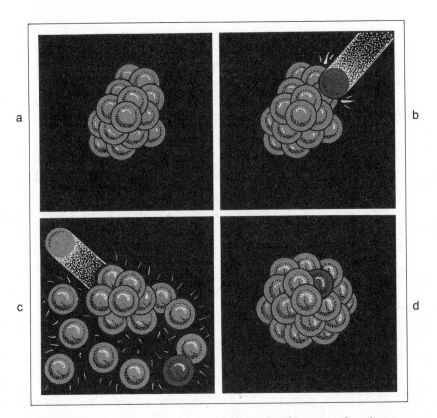

a

b

c

d

The alpha particle with a mass number of 4 (b) enters the nitrogen nucleus (a & b), and a proton with a mass number of 1 is expelled (c). The net result is that nitrogen-14 (a) combines with helium-2 (an alpha particle) (b), to throw off hydrogen-1 (a proton) (c), and produces oxygen-17 (d).

with its mass number of 4, and a proton had left with its mass number of 1. The nitrogen nucleus had gained 3 in its mass number, increasing from 14 to 17. The net result was, then, that nitrogen-14 had combined with helium-2 (an alpha particle) to yield oxygen-17 and hydrogen-1 (a proton).

Rutherford, therefore, was the first to bring about a nuclear reaction in the laboratory; that is, he was the first to bring about the change of one element into another—

nitrogen into oxygen—through human agency. For the use of the cloud chamber in this case and others, Blackett received a Nobel prize in 1948.

In a way, Rutherford had brought about what the old alchemists would have called the transmutation of elements, and there were some people who, upon hearing of his work, said, "See, the old alchemists were right, after all. Modern scientists were wrong to dismiss them with scorn." This view is wrong, however. The alchemists did not merely maintain that transmutation was possible, but thought it could be brought about by chemical means alone—by mixing, heating, distilling, and so on. In this they were wrong. Transmutation can be brought about only by nuclear reactions, which were beyond anything possible to the old alchemists and beyond anything they had any idea of.

In short, an idea is not enough. Significant details must also be correct before one can get credit for "being right." Thus, there were people before Newton who talked of trips to the Moon, and, in a way, this was, in itself, a sensible idea. It was Newton, however, who first showed that a trip to the Moon could only be achieved by means of the rocket principle. It was he, not his predecessors, who should get the credit—not for a mere dream, but for a dream that included a practical pathway to its fulfillment.

Artificial Isotopes

Rutherford had changed one isotope known in nature, nitrogen-14, into another known isotope, oxygen-17. After the possibility of such laboratory transmutations had been established, other nuclear reactions, producing other

164

known isotopes, were brought about by bombarding various types of atoms with speeding particles.

But need the changes always produce isotopes that were already known? Might not adding and subtracting particles produce nuclei with mass numbers and charges not quite like any that occurred naturally? In 1932, the Latvian-American chemist Aristid V. Grosse (b. 1905) suggested that it might be possible to do this.

In 1934, the Joliot-Curies were continuing Rutherford's work on the bombardment of various elements by alpha particles. They were bombarding aluminum in this fashion and not only knocked protons out of the aluminum nuclei, but in some cases knocked neutrons out of them, too. When the bombardment was over, the streams of protons and neutrons that emerged from the aluminum nuclei stopped at once. To their astonishment, however, some sort of radiation (of a type that will be taken up later in the book) continued, and declined with time just as the radiation intensity of a radioactive substance might be expected to decline. They could even calculate the half-life of the radiation as being 2.6 minutes.

All aluminum atoms in nature have an atomic number of 13 and a mass number of 27. In other words, their nuclei are all composed of 13 protons and 14 neutrons. If you add an alpha particle (2 protons and 2 neutrons) and knock out a proton, the new nucleus contains 14 protons and 16 neutrons, which is silicon-30, a well-known isotope.

But what about the cases in which a neutron is knocked out of the nucleus? If, to an aluminum nucleus (13 protons and 14 neutrons) you add an alpha particle (2 protons and 2 neutrons) and knock out a neutron, you end up with a new nucleus consisting of 15 protons and 15 neutrons. This is phosphorus-30. Phosphorus-30 does not, however, occur in nature. All phosphorus atoms in nature are phosphorus-31 (15 protons and 16 neutrons), which is the only stable phosphorus isotope. Phosphorus-30 is radioactive and quickly breaks down (by a method we'll describe later) into the stable silicon-30.

Phosphorus-30 was the first "artificial" isotope produced, and it introduced the concept of artificial radioactivity. The Joliot-Curies shared a Nobel prize in 1935 for this work.

After the Joliot-Curies had shown the way, a great many different artificial isotopes were formed through nuclear reactions of one sort or another. Every one of them was radioactive, and so one spoke of radioactive isotopes or radioisotopes.

All of the stable isotopes or nearly stable isotopes that exist are to be found in Earth's rocks. None of the radioisotopes formed in the laboratory, so far, have half-lives long enough to make it possible for any measurable quantity of them to have endured since the Earth's beginnings.

Every known element has radioisotopes. Even hydrogen, the simplest, has a radioactive isotope, hydrogen-3, the nucleus of which consists of 1 proton and 2 neutrons. It is sometimes called tritium, from a Greek word for "third." It has a half-life of 12.26 years. Tritium was first prepared in the laboratory in 1934 by the Australian physicist Marcus Laurence Elwin Oliphant (b. 1901).

For a quarter century after Rutherford's pioneering work, scientists kept pounding away at atoms, with alpha particles as their projectiles. This had its advantages. For one thing, alpha particles were always available. Uranium, thorium, and several of their breakdown products (radium, for instance) produced alpha particles, therefore there would always be a supply of them.

There were disadvantages, too. Alpha particles were positively charged, as were atomic nuclei. (After all, the alpha particle is itself an atomic nucleus.) This meant that the nuclei repelled the alpha particles, and before an alpha

166

particle could collide with and enter an atomic nucleus, that alpha particle had to overcome the repulsion. Some of its energies were consumed in doing so, and that reduced its effectiveness. Aditionally, the more massive the nucleus being bombarded, the greater the repulsion. Beyond a certain point, available alpha particles could not enter a nucleus at all.

Once the neutron was discovered, however, Enrico Fermi realized that here was a unique new projectile. If a stream of neutrons was produced, let us say, by having a stream of protons strike paraffin, these neutrons, being uncharged, were not repelled by atomic nuclei. If a neutron happened to be moving in the direction of a nucleus, it could strike it and enter it even if it had very little energy. The discovery of the neutron thus revolutionized the entire technique of atom bombardment.

Fermi found that if he sent a stream of neutrons through water or paraffin, many of the neutrons would strike nuclei but bounce off without penetrating, losing some energy in the process. Eventually, these neutrons would have only the energy expected of something jiggling along with the usual speed that particles have at a given temperature. These would become thermal neutrons or slow neutrons. Such slow neutrons, Fermi found, were actually more likely to be absorbed by nuclei than were fast neutrons.

Fermi also found that when a neutron entered a nucleus, a beta particle (an electron) was usually emitted. The addition of the neutron raised the nucleus's mass number by 1, and the ejection of a beta particle, by subtracting 1 negative charge, raised the nuclear charge (that is, the atomic number) by 1. In short, neutron bombardment of a particular element tended to produce the next higher element in the atomic number scale.

In 1934, it occurred to Fermi that it might be very

167

interesting to bombard uranium with neutrons. Uranium, with an atomic number of 92, had the highest atomic number known at that time. If it were bombarded with neutrons, and made to emit beta particles, would it not form element 93, which was unknown in nature?

Fermi tried the experiment, and it seemed to him that he did indeed get element 93. The experiment, however, yielded complex and confusing results, and (as we shall see later) it took some years to straighten out the findings.

The Italian physicist Emilio Segrè (1905–1989), who had worked with Fermi, decided it was not necessary to bombard uranium with neutrons in order to create an unknown element. At this time, the mid-1930s, there were four spaces in the periodic table that remained unfilled. These represented unknown elements. Of these, the one with the lowest atomic number was element 43.

In 1925, a group of German chemists, including Walter Karl Friedrich Noddack (1893–1960) and Ida Eva Tacke (b. 1896), reported the discovery of element 75, which they called rhenium after the Latin name for the Rhine in Germany. It turned out to be the last of the 81 stable elements to be discovered. The group also announced that they had found traces of element 43, which they called masurium after a region in eastern Germany.

The second announcement, however, turned out to be a false alarm, and element 43 remained undiscovered. Why not bombard molybdenum (element 42) with neutrons, thought Segrè, to see if element 43 could at least be manufactured, if not found.

In 1937, Segrè went to the United States to have molybdenum bombarded with neutrons by a new technique (which I will describe later), and he did indeed locate element 43 in the bombarded material. Segrè hesitated to name the new element, however, because he wasn't certain that an artificially produced element was the equivalent of

168

finding one in nature. In 1947, however, the German-British chemist Friedrich Adolf Paneth (1887–1958) maintained vigorously that it *was* the equivalent, and this view was accepted. Segrè therefore named element 43 technetium, from a Greek word meaning "artificial."

Enough technetium was formed one way or another for its properties to be studied, and it was found that three of its isotopes were quite long-lived. The longest-lived is technetium-97 (with a nucleus containing 43 protons and 55 neutrons), which has a half-life of 2,600,000 years. On a human scale, a sample of this isotope would seem permanent; only a tiny bit of the material would have broken down in the course of a human lifetime. Nevertheless, there are *no* stable isotopes of technetium, and even the most nearly stable one, technetium-97, is not long-lived enough to have persisted since Earth was formed. Even if there had been large quantities in the soil in Earth's youth, none would now remain. This is especially so because no isotope of technetium is formed from any other, longer-lived radioactive element.

The remaining three vacancies in the periodic table at this time were elements 61, 85, and 87. All three had been occasionally reported as having been discovered in minerals of one sort or another, but all of the reports turned out to be mistaken.

In 1947, however, the American chemist Charles D. Coryell (b. 1912) and his co-workers located element 61 in the products of uranium breakdown after neutron bombardment (something we are going to return to). They named it promethium after the Greek god Prometheus, who brought fire from the Sun to humanity, because the element had been found in the Sun-like fire of a nuclear reaction. None of the isotopes of promethium are stable, and even the longest-lived, promethium-145 (61 protons and 84 neutrons), has a half-life of only 17.7 years.

In 1939, the French chemist Marguerite Perey (b. 1909) located tiny traces of element 87, as a very minor breakdown product of uranium-235. She named the element francium for France. The isotope she had located was francium-215 (87 protons, 128 neutrons). Its half-life is only slightly over a millionth of a second, so Perey certainly didn't detect the isotope itself. What she did detect were the very energetic alpha particles it produced (the shorter the half-life of an alpha-particle producer, the more energetic the alpha particles), and was able to reason from what was already known about the breakdown pattern the isotope that had to be responsible. Even the most long-lived francium isotope, francium-223 (87 protons, 136 neutrons) has a half-life of only 21.8 minutes.

In 1940, element 85 was produced, by Segrè and others, by the bombardment of bismuth (element 83) with alpha particles. Element 85 was named astatine, from Greek words meaning "unstable," because, like all other elements discovered since 1925, it was exactly that. Its longest-lived isotope, astatine-210 (85 protons, 125 neutrons), has a half-life of 8.1 hours.

By 1948, then, the periodic table had been filled from hydrogen (1) to uranium (92), and elements beyond uranium had been discovered. Fermi thought he had produced, in 1934, element 93 in his bombardment of uranium by neutrons, but it was not until 1940 that the element was actually isolated in bombarded uranium by the American physicists Edwin Mattison McMillan (b. 1907) and Philip Hauge Abelson (b. 1913). In that uranium had been named for the then newly discovered planet Uranus, McMillan named element 93, which lay just beyond uranium, neptunium, after Neptune, the planet that lay beyond Uranus.

Neptunium-237 is the longest-lived isotope of that element, having a half-life of 2,140,000 years. This is long, but not nearly long enough to allow any neptunium to remain in the Earth's crust, even if there had been quite a bit there to begin with. Nevertheless, neptunium-237 has its interest because it breaks down through a series of intermediate compounds, as do uranium-238, uranium-235, and thorium-232.

In fact, neptunium-237 initiates the fourth radioactive series referred to earlier. It and all of its breakdown products have mass numbers that are divisible by 4, with a remainder of 1. There are only four radioactive series possible in these upper reaches of the periodic table: thorium-232 (remainder 0), neptunium-237 (remainder 1), uranium-238 (remainder 2), and uranium-235 (remainder 3). Of these, three exist today, but neptunium-237 is extinct, for even the longest-lived member of the series didn't have a half-life long-lived enough for the isotope to be around now.

Another odd thing about the neptunium-237 series is that it is the only one that doesn't end in a stable isotope of lead. It ends with bismuth-209, the only stable isotope of bismuth.

In 1940, the American physicist Glenn Theodore Seaborg (b. 1912) joined McMillan to find that certain neptunium isotopes give off beta particles and become isotopes of the same mass number, but with an atomic number higher by 1. Thus, they discovered element 94, naming it plutonium after Pluto, the planet beyond Neptune. Its most long-lived isotope is plutonium-244 (94 protons, 150 neutrons), with a half-life of 82,000,000 years. McMillan and Seaborg shared a Nobel prize in 1951 for finding a transuranian (i.e., "beyond uranium") element.

McMillan went on to other endeavors, but Seaborg and others continued to produce additional elements. The following transplutonian elements have been isolated:

Americium (for America), with an atomic number of 95. Its longest-lived isotope is americium-243 (95 protons, 148 neutrons), with a half-life of 7,370 years.

171

Curium (for the Curies), with an atomic number of 96. Its longest-lived isotope is curium-247 (96 protons, 151 neutrons), with a half-life of 15,600,000 years.

Berkelium (for Berkeley, California, where it was discovered), with an atomic number of 97. Its longest-lived isotope is berkelium-247 (97 protons, 150 neutrons), with a half-life of 1,400 years.

Californium (for California, the state in which it was discovered), with an atomic number of 98. Its longest-lived isotope is californium-251 (98 protons, 153 neutrons), with a half-life of 890 years.

Einsteinium (for Albert Einstein), with an atomic number of 99. Its longest-lived isotope is einsteinium-252 (99 protons, 153 neutrons), with a half-life of 1.29 years.

Fermium (for Enrico Fermi), with an atomic number of 100. Its longest-lived isotope is fermium-257 (100 protons, 157 neutrons), with a half-life of 100.5 days.

Mendelevium (for Dmitri Mendeleev), with an atomic number of 101. Its longest-lived isotope is mendelevium-258 (101 protons, 157 neutrons), with a half-life of 56 days.

Nobelium (for Alfred Nobel, who established the Nobel prizes), with an atomic number of 102. Its longest-lived isotope, so far detected, is nobelium-259 (102 protons, 157 neutrons), with a half-life of about 58 minutes.

Lawrencium (for Ernest Lawrence, mentioned later in the book), with an atomic number of 103. Its longest-lived isotope, so far detected, is lawrencium-260 (103 protons, 157 neutrons), with a half-life of 3 minutes.

Rutherfordium (for Ernest Rutherford), with an atomic number of 104. Its longest-lived isotope, so far detected, is rutherfordium-261 (104 protons, 157 neutrons), with a half-life of 65 seconds.

Hahnium (for Otto Hahn), with an atomic number of 105. Its longest-lived isotope, so far detected, is hahnium-262 (105 protons, 157 neutrons), with a half-life of 34 seconds.

Element 106 has been found, but two groups claim the discovery. The case hasn't been settled, and until it is, there can be no official name. The longest-lived isotope so far detected has a mass number of 263 (106 protons, 157 neutrons), with a half-life of 0.8 second.

It is uncertain how much further scientists can go. As the atomic numbers go up, the elements are harder to produce and, because the half-lives tend to decrease, harder to study. There is considerable pressure to reach elements 110 and 114, however, for there are compelling arguments in favor of supposing that some isotopes of these elements are long-lived or even stable.

7

BREAKDOWNS

Mass Defect

As previously noted, the current standard used in atomic weight measurements is carbon-12. The mass number of carbon-12 is defined as 12.0000, and all other mass numbers are measured against this number. Atomic weights, which are the weighted averages of the mass numbers of the isotopes making up a particular element, are also measured against the carbon-12 standard.

Carbon-12 has twelve particles in its nucleus: 6 protons and 6 neutrons. On average, each of the particles should, then, have a mass of 1.0000 if all twelve constitute mass 12.0000. However, modern mass spectrographs, capable of measuring the mass of individual protons curving in a magnetic field of known intensity, show the mass of the proton to be not 1.0000, but 1.00734.

A neutron, being uncharged, does not curve in passing through a magnetic field, but its mass can be worked out in other ways. In 1934, Chadwick measured the exact amount of energy it took to break apart the proton and neutron in a hydrogen-2 nucleus. The mass of a hydrogen-2 nucleus is known. From that, subtract the mass of the proton and add the mass of the energy used to break it up (calculated from Einstein's equation relating mass and energy). What is left over is the mass of the neutron.

It turns out that the mass of the neutron is 1.00867. In other words, the proton and neutron are not exactly equal in mass. The neutron is about 1/7 of 1 percent more massive than the proton (which, as we shall see, is important).

If we imagine ourselves taking 6 protons and 6 neutrons, considering them as separate particles, and adding up their individual masses, the total mass is 12.096. If, however, we squeeze all 6 protons and 6 neutrons tightly together into a carbon-12 nucleus, we have a total mass of 12.0000.

The mass of the carbon-12 nucleus is 0.096 less than it ought to be if the masses of the individual particles making it up are added together. In 1927, Aston, working with his mass spectrograph, found that all nuclei had masses a trifle less than would be expected if the masses of their separate particles were added together. Aston referred to this as a *mass defect.*

What fraction of the total mass of carbon-12 is the mass defect of this nucleus? The fraction is 0.096 divided by 12, or 0.008. In order to avoid working with this small decimal, scientists multiply it by 10,000. This gives us 80, which is the *packing fraction* of carbon-12.

In general, if we start with hydrogen-1 and its single-

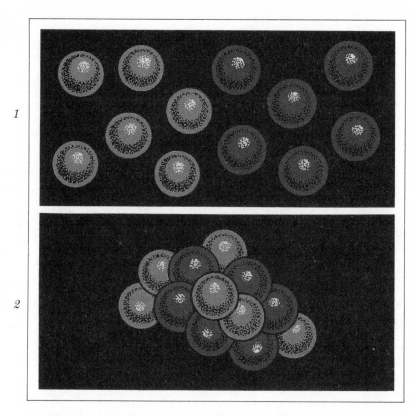

1. *Carbon 12's nucleus is composed of 6 protons and 6 neutrons. Their individual masses add up to 12.096.*
2. *Squeezed together into a nucleus, the total mass is only 12.000. This phenomenon is called the* mass defect.

proton nucleus and go up the scale of stable isotopes, we find that the packing fraction gets greater and greater, until we get to iron-56. Iron-56 has a nucleus containing 26 protons and 30 neutrons. If these particles are considered separately, the total mass is 56.4509. The mass of the iron-56 nucleus, however, is measured as 55.9349. The mass defect is 0.5260 (about half a proton). The packing fraction, 0.5260 divided by 55.9349 multiplied by 10,000, comes out to 94.0. If we continue going up the scale of stable isotopes that follow iron-56, we find that the packing fraction begins to decrease again. By the time we reach uranium-238, the packing fraction is only 79.4.

What happens to the mass that disappears when protons and neutrons are packed into a nucleus? There is only one thing that can happen: it turns into energy in accordance with Einstein's equation. In other words, if a carbon-12 nucleus is formed out of 6 protons and 6 neutrons, a small fraction of the mass of these particles is turned into energy, which dissipates into surrounding space. Any process that dissipates energy has a tendency to take place spontaneously (although not always quickly). This means that there is a tendency, under proper conditions, for protons and neutrons to combine to form nuclei.

On the other hand, in order to break up an atomic nucleus into its individual protons and neutrons, an amount of energy must be supplied that is exactly equal to the energy dissipated in the formation of that nucleus. But how can dissipated energy be collected once more and crowded into the tiny volume of a nucleus. Except under the most extraordinary conditions, this will not happen; nuclei do *not* tend to break up into their individual protons and neutrons. Once nuclei have formed, they tend to retain their identity indefinitely.

A nucleus, however, need not blow apart into single particles in losing its identity. What if it simply gains or loses a proton, or a neutron? Such an event would be sufficient to change one nucleus into another with, possibly, a further dissipation of energy. On the whole, if one nucleus changes into another nucleus with a higher packing fraction, there is a further dissipation of energy and there is a certain tendency for that change to take place.

We might, then, expect that nuclei with very low mass numbers would tend to change into those of higher mass numbers, while nuclei with very high mass numbers would tend to change into those of lower mass numbers. The nuclei at both extremes of the mass number scale would tend to converge at iron-56. Iron-56, with its maximum packing fraction, would require an energy input to become either larger or smaller in order for it to stay put.

A tendency, however, need not be observed in fact. If we are standing on sloping land, we have a tendency to slide downhill. If, however, the land is rough and uneven and we are wearing sneakers with ridged rubber soles, the amount of friction set up would prevent our sliding down, despite the tendency. On the other hand, if the slope were made steeper, or if it remained unchanged but grew icy, the friction might not suffice to keep us put and we would again tend to slide downhill.

To take another example, paper has a tendency to burn; that is, to combine with the oxygen of the atmosphere. Nevertheless, the chemical changes required to produce the effect of burning do not take place at ordinary temperatures because an energy of activation is required. (This is a type of friction that prevents a chemical change that "should" take place from actually taking place.) If the paper is heated, however, more and more energy enters it and, eventually, the value of the energy activation is topped, with the paper bursting into flame.

Once the paper flames, enough heat is produced to serve as an energy of activation for neighboring regions of the paper, which flame in their turn, spreading the flame still farther. The paper can thus continue burning indefinitely without further supplies of energy from outside. The small initial supply is all one needs. The proverbial smoldering cigarette butt that eventually burns down an entire forest is a very realistic scenario. This process, whereby a chemical reaction produces what is needed to initiate a further installment of that same chemical reaction, is called, simply, a chain reaction.

In the case of nuclei, there are factors that prevent the natural tendency of sliding toward iron-56 from turning into actuality. This is particularly true of the light nuclei, for reasons we will go into later.

For the massive nuclei, the tendency is more easily realized. Indeed, for all known nuclei that are more massive than that of bismuth-209, the tendency *is* realized. The more massive nuclei tend to emit particles in such a way that the newly formed nucleus is smaller and, therefore, has a higher packing fraction than the old. In this way, energy is produced and dissipated.

The greater the dissipation of heat in the change from one nucleus to another, the more likely is the change, the more rapidly the change takes place, and the shorter the half-life of the original nucleus. In the cases of thorium-232, uranium-235, and uranium-238, the initial change involves so small a dissipation of heat that their half-lives are very long. Nevertheless, even here the half-lives are not infinite and the change does take place, but slowly. (By way of analogy, although paper doesn't burst into flame at ordinary temperatures, it does undergo changes very slowly. Every once in a while a chemical change can take place even though the energy of activation requirement is not met. Thus, as the years pass, the pages of a book often slowly turn browner and more brittle until they eventually crumble into an ash that is the result of very slow "burning." We might say that the paper molecules have a "burn half-life" at room temperature, which might be long from the standpoint of a reader, but which is much shorter than that of uranium-238.)

To put it briefly, the type of natural radioactivity discovered in the 1890s changes nuclei with mass numbers 232 to 238 into nuclei with mass numbers 206 to 208. In the process, the packing fraction increases, whereby the protons and neutrons in the final nuclei have slightly less mass than those in the original nuclei. The missing mass is dissipated as energy, which is how we explain where the energy produced by radioactivity comes from.

Nuclear Fission

I mentioned earlier in the book that Enrico Fermi, in 1934, had bombarded uranium with slow neutrons in order to form element 93 (definitively located six years later and named neptunium). Fermi thought he had located the element—and, in a way, he had—but studies of the bombarded uranium seemed to show such a confusing melange of particles that it was difficult to precisely locate element 93 in the mixture. (Still, Fermi received a Nobel prize in 1938 for this work.)

Ida Tacke (one of the codiscoverers of the element rhenium) suspected that what had happened in Fermi's work was that the uranium nucleus was so complex and, therefore, so rickety, that on absorption of a neutron the nucleus simply fell apart into fragments. This was so different from anything that had been observed in nuclear breakdowns, however, that no one paid much attention to the suggestion.

Beginning in 1937, however, the team of Hahn and Meitner tackled the problem in Germany. (Meitner was Jewish, but she was an Austrian national, so she was momentarily safe from Adolf Hitler, 1889–1945—who then ruled Germany—and his malignant anti-Semitic policies.)

Hahn decided that what might have happened in the case of the neutron bombardment of uranium was that the uranium lost *two* alpha particles rather than one. That was as far as he dared go in the direction of Tacke's notion of fragmentation. The loss of two alpha particles would reduce

180

the atomic number by 4, from uranium's 92 to radium's 88. If Hahn were correct, then, there ought to be slightly higher quantities of radium in the bombarded uranium than the usual traces to be expected from ordinary radioactive changes.

How could one detect these minute traces of radium and estimate the quantity present? Marie Curie had isolated traces of radium from uranium ore, but she had had tons of the ore to work with. Hahn and Meitner had only a small quantity of bombarded uranium.

As it happens, radium, in the periodic table, is placed just under the stable element barium, the two elements being very similar in chemical properties. If bombarded uranium is dissolved in acid and barium is added to the solution, the barium can be separated out again by simple chemical procedures and the radium will come out with it. (The radium will do whatever the barium does.)

If, then, Hahn and Meitner put in perfectly stable barium and got out distinctly radioactive barium, they would know that radium had been extracted with the barium. From the amount of radioactivity present (easily measured) they could determine, delicately, the amount of radium obtained. Before this experiment could be carried through, however, Nazi Germany invaded Austria, annexing it in March of 1938. Meitner then became subject to Hitler's anti-Semitism, whereupon she slipped across the border to the Netherlands, and from there to Stockholm, Sweden.

Hahn continued to work, with the German chemist Fritz Strassman (1902–1980). They added stable barium to solution and got out radioactive barium, from which they could estimate the quantity of radium they had extracted. Then, in order to produce the final step of the proof, it was necessary to separate the radium from the barium in order to produce a solution containing the radium alone.

That separation, however, didn't work. Nothing Hahn

and Strassman tried would separate the radium from the barium. Hahn decided that if nothing would separate the stable barium from the radioactive radium, the radioactive atoms were *not* radium, but barium—or more correct, a radioisotope of barium. (Hahn received a Nobel prize in 1944 for this insight.)

But how could uranium break down to yield barium? The atomic number of barium was 56. If uranium (atomic number 92) had split into barium, it would have to give off 18 alpha particles, or it would have to split in two. Either alternative seemed so unlikely that Hahn didn't quite dare suggest them publicly.

At the same time, in Stockholm, Meitner was coming to precisely the same conclusion when she received news of the failure to separate the supposed radium from barium. She *did* decide to go public. With the help of her nephew, the physicist Otto Robert Frisch (1904–1979), Meitner prepared a letter, dated January 16, 1939, and sent it to the British journal *Nature*. Frisch, who was working in Bohr's laboratory in Copenhagen, told Bohr of the contents of the letter before it was published. Bohr went to the United States to attend a physics conference in Washington, D.C., on January 26, 1939, and there he spread the word—also before the letter was published.

Concurrently, in Great Britain, the Hungarian physicist Leo Szilard (1898–1964)—who, like Meitner, had fled Germany because he was Jewish—was thinking about what H. G. Wells termed, in his science fiction, an atomic bomb. Szilard thought that such a bomb might exist if a neutron were to strike an atomic nucleus, thus bringing about a change that would cause the ejection of two neutrons, which would strike two nuclei and bring about the ejection of four neutrons, and so on. The number of breakdowns per second and the energy liberated would quickly rise to enormous proportions, thereby producing a vast explosion. Szilard was, in fact, visualizing a nuclear chain reaction.

Szilard even patented the process. He enlisted the help of another Jew, the Russian-British biochemist Chaim Weizmann (1874–1952), who tried to carry through the necessary experimentation. However, it failed. Nuclei that absorbed and ejected neutrons absorbed only fast, energetic ones, and ejected slower ones that were too lacking in energy to keep the reaction going.

But then Szilard heard of the shattering of the uranium nucleus as a result of the absorption of a neutron. (This break of a nucleus into two nearly equal parts came to be called fission, from a Latin word meaning "to split." We often speak of uranium fission, but uranium nuclei are not the only ones that undergo such a split; therefore, it is better to use the more general term nuclear fission.)

Szilard saw at once that here was the possibility of a practical version of the nuclear chain reaction he had visualized. It was a slow neutron that split the uranium nucleus and, in the process, it was soon discovered that two or three slow neutrons were liberated for each nucleus that split.

In 1940, Szilard labored to persuade American physicists to establish self-censorship over their investigations into nuclear fission, for fear that German physicists might benefit from American findings and give Adolf Hitler a devastating new type of bomb. (World War II had already started, with Germany making successful advances and the United States still neutral.)

Szilard next had to persuade the United States government to pour enormous funds into the research. He obtained the help of two other Hungarian-born refugees from Nazism: Eugene Paul Wigner (b. 1902) and Edward Teller (b. 1908). In 1942, the three visited Albert Einstein, still another such refugee. Einstein, as the only scientist whose word would carry sufficient weight, agreed to write a letter to President Franklin Delano Roosevelt (1882–1945). Roosevelt received the letter, considered it, was persuaded by it, and, on a Saturday late in the year, signed the order

that set up what came to be called the Manhattan Project—
a deliberately meaningless name designed to mask its real
purpose.

As it happened, the timing of Szilard's push for the
project was precarious. It is not often that work is done on
a weekend. If Roosevelt had delayed the signing until Mon-
day, chances are that it would not have been signed at all—
for the Saturday he signed it was December 6, 1941, and
the next day Japan attacked Pearl Harbor. Who can tell
when Roosevelt would have been able to think of the order
again. At any rate, the project proceeded and an atomic
bomb (more properly, a nuclear fission bomb) was developed by July 1945, after Germany was crushed and Hitler
had committed suicide. It was used to finish off a virtually
helpless Japan, on August 6 and 8, 1945.

When a uranium nucleus undergoes fission, it does not
always divide in exactly the same way. The packing fraction
among nuclei of moderate size does not vary a great deal,
and the uranium nucleus may well break at one point in
one case and at a slightly different point in another. For
this reason, a mixture of a great variety of radioisotopes
are produced during uranium fission. They are lumped together under the term fission products.

The probabilities are highest that the division will be
slightly unequal, with a more massive part in the mass
region from 135 to 145, and a less massive part in the region
from 90 to 100. It was among the more massive part that
the element promethium was located in 1948.

As a result of nuclear fission, the uranium nucleus
slides farther down the packing fraction hill than it does in
its ordinary radioactive breakdown to lead. Uranium fission, therefore, releases considerably more energy than
ordinary uranium radioactivity. (Nuclear fission can also
release it much more quickly if a chain reaction is set up,
pouring forth its energy content in a fraction of a second

where ordinary uranium radioactivity would take billions of years.)

But if more energy is released in nuclear fission than in ordinary radioactivity, why doesn't the uranium nucleus break down by fission naturally, instead of giving off a series of alpha and beta particles? The answer is that there is a higher energy of activation involved in the nuclear fission process. The necessary energy of activation can be supplied if a neutron floats into a nucleus, changes its nature, and sets it vibrating, but not otherwise. At least, *almost* not otherwise. Despite the higher energy of activation, a uranium nucleus will very occasionally undergo spontaneous fission, slipping through the wall, so to speak, set up by the energy of activation. It happens very rarely, however. A uranium-238 nucleus undergoes 1 spontaneous fission for every 220 times such nuclei simply give up alpha particles. Such spontaneous fission was first detected in 1941 by the Soviet physicist Georgii Nikolaevich Flerov (b. 1913).

Just as some radioactive nuclei have much shorter half-lives than others, so do some show much greater likelihood of undergoing spontaneous fission. The transuranic isotopes, for instance, become more unstable not only with regard to ordinary radioactivity but with regard to spontaneous fission as well. Where the spontaneous fission half-life of uranium-238 is about one trillion years, that of curium-242 is 7,200,000 years, and that of californium-250 is only 15,000 years.

Uranium-238 undergoes fission with such difficulty that even neutron bombardment is insufficient to bring about much of it. It takes fast, energetic neutrons to do the trick, while only slow neutrons are ejected so that no chain reaction is possible.

It was Bohr who pointed out, soon after fission was established, that, on theoretical grounds, it was uranium-235 that should set up the chain reaction. Uranium-235 is less stable than uranium-238. Uranium-235 has a half-life only 1/6 that of uranium-238, and even a slow neutron will cause it to undergo fission. One of the more difficult aspects of developing the fission bomb, in fact, was separating uranium-235 from uranium-238, since ordinary uranium, as found in nature, simply doesn't have enough uranium-235 in it to support a nuclear chain reaction.

It is possible, however, to bombard uranium-238 with neutrons in such a way as to form first neptunium-239 and then plutonium-239. Plutonium-239 has a half-life of over 24,000 years, which is long enough to allow it to be accumulated in quantity. Plutonium-239 is fissionable with slow neutrons, as is uranium-235.

Again, thorium-232 is not fissionable with slow neutrons. When thorium-232 is bombarded with neutrons, however, it can become thorium-233, which, in turn, becomes uranium-233. Uranium-233 was first discovered by Seaborg in 1942. It has a half-life of 160,000 years and is fissionable by slow neutrons.

In other words, all of the uranium and thorium in the world can, in theory, be converted into nuclei that can be made to undergo fission, and, if controlled, to yield useful energy instead of merely a wild explosion. Thorium and uranium are not very common elements, but taken together, there is ten times as much energy available from them as from the Earth's entire supply of coal, oil, and gas.

In the 1950s, nuclear reactors yielding controlled energy began to be built, and a significant portion of the world's energy is now derived from them. To be sure, safety is a concern. (The Three-Mile Island accident in the United States in 1979 and the true disaster at Chernobyl in the Soviet Union in 1986 caused considerable alarm.) Then, too, there is a question of the disposal of the ever-accumulating fission products, which are dangerously radioactive. For these reasons, the future of fission energy seems cloudy at the moment. There is, however, another type of nuclear energy that might prove just as useful, while inherently safer to use than fission.

Nuclear Fusion

Both natural radioactivity and nuclear fission affect nuclei with large mass numbers, bringing about a change toward nuclei of intermediate mass numbers and greater stability. In the process, mass is lost and energy is produced and dissipated. It is also possible for nuclei with small mass numbers to combine with each other, or fuse (that is, melt together), forming a somewhat more massive nucleus. This, too, represents a change toward nuclei of intermediate mass numbers and greater stability. Here, too, mass is lost and energy is produced and dissipated.

Indeed, whereas the packing fraction, as one goes from large mass numbers to intermediate ones, is marked by a rather gentle rise; the packing fraction, as one goes from small mass numbers to intermediate ones rises much more steeply. This means that nuclear fusion can produce more energy from a given mass of starting material than can nuclear fission.

Let's examine how this works by taking the example of a nucleus of hydrogen-2 (1 proton plus 1 neutron) fusing with another nucleus of hydrogen-2 to form a nucleus of helium-4 (2 protons plus 2 neutrons). The mass number of hydrogen-2 is 2.0140, and two of them are 4.0280. Helium-4, however, has an unusually high packing fraction for its size, whereby its mass number is only 4.0026. The loss of mass number in going from two hydrogen-2 nuclei to a helium-4 nucleus is $4.0280 - 4.0026 = 0.0254$. The loss in mass of 0.0254 is 0.63 percent of the original mass of 4.028.

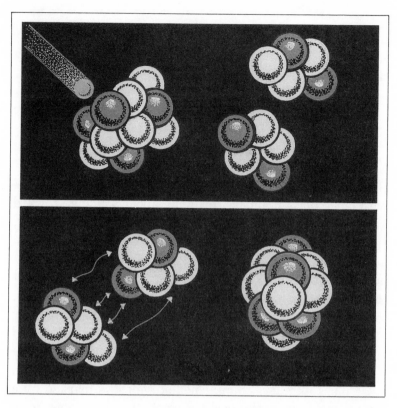

Both fission, the splitting of an atom (top), and fusion, the joining together of nuclei (bottom), liberate energy. Kilogram for kilogram, the fusion of hydrogen can produce eleven times as much energy as can uranium fission.

This doesn't sound like much (hydrogen-2 in fusing to helium-4 loses only ⅝ of 1 percent of its mass), but it is actually a great deal. The natural radioactive change of uranium-238 to lead-206 results in the loss of only 0.026 percent of the original mass, whereas the nuclear fission of uranium-235 results in the loss of only 0.056 percent of its original mass. The fusion of hydrogen can produce about twenty-four times as much energy, kilogram for kilogram, as can natural radioactivity, and 11 times as much as can uranium fission.

The energy that could be produced by nuclear fusion proved of crucial importance to an understanding of the Universe even before that other energy source, nuclear fission, was discovered. The story goes as follows.

Ever since the development of the law of conservation of energy in 1847, scientists had been wondering as to the source of the energy radiated by the Sun. This radiation had been occurring throughout human history, and geological studies made it plain that it had been occurring long before human beings had evolved.

No source of energy known in the nineteenth century could account for the Sun's having kept burning for more than a hundred million years or so, which proved to be a conservative estimate. In the first decade of the twentieth century, scientists began measuring the ages of rocks and of meteorites by the amount of radioactive change that had taken place in them. It soon appeared that our solar system (including the Earth and the Sun) was several billion years old. The best current figure for its age is 4,550,000,000 years.

By 1910, it was recognized that nuclear energy was more powerful than any other that had been considered. In 1920, the British astronomer Arthur Stanley Eddington (1882–1944) suggested that the Sun's energy might be derived from the fusion of hydrogen to helium. This suggestion looked better as the decade progressed. When Aston worked out the notion of packing fraction, it became clear that the fusion of hydrogen to helium was the only simple nuclear reaction that could possibly yield enough energy to power the Sun.

Then, in 1929, the American astronomer Henry Norris Russell (1877–1957) worked out the chemical composition of the Sun by a careful study of its light spectrum, and discovered that it was mostly hydrogen. About 90 percent of the atoms in the Sun are hydrogen, and 9 percent are helium. All of the other elements together make up the remaining 1 percent of the atoms. This meant that not only was the fusion of hydrogen to helium the only simple nuclear reaction that could yield enough energy, it was the only significant nuclear reaction that could take place at all. It was hydrogen fusion or nothing.

In 1938, the German-American physicist Hans Albrecht Bethe (b. 1906) worked out the details of what must go on in the center of the Sun, basing his theory on what was known from laboratory studies of nuclear reactions and on astronomical inferences as to conditions at the center of the Sun. He received a Nobel prize in 1967 for this work.

It is now thought that most normal stars are constantly fusing hydrogen, which can serve as an energy source for billions of years. Eventually, especially in the case of the more massive stars, conditions at the center of a star become such that helium nuclei are further fused into still more massive nuclei, such as those of carbon, oxygen, neon, silicon, through to iron (where the process stops because the packing fraction has reached its maximum).

Very massive stars that have gone as far as they can in the process of fusion find their energy source failing and can no longer support the weight of their own outer layers. The star collapses and, in the process, all of the hydrogen remaining in the outer layers (together with other atoms of mass number smaller than iron) fuses at once. The result is a vast outpouring of energy—an enormous explosion we call a supernova. Much of the material of the exploding star is spewed into surrounding space by the explosion, while what remains collapses into a tiny object called a neutron star, or into an even tinier object called a black hole.

It is now thought that at the time the Universe was first formed only hydrogen and helium nuclei were created. The more massive nuclei were formed only in the centers of stars, and it was only because some massive stars exploded that these more massive nuclei were added to the dust, gas, and debris of material in space. In fact, as a supernova explodes, so much energy becomes available that iron nuclei are driven "uphill" to form even more massive nuclei all the way up to uranium and beyond—and these, too, pour into surrounding space.

Eventually, new stars form out of the interstellar dust and gas containing the massive nuclei that had been spread outward by supernova explosions. These new stars are second-generation stars. They and their planets contain great quantities of the massive nuclei.

The Sun is such a second-generation star. The Earth, and we, ourselves, are made almost entirely of massive nuclei that were once formed at the center of a giant star and then spread through space by a giant explosion.

But if hydrogen fusion produces so much more energy than natural radioactivity does, why doesn't it take place spontaneously, and much more rapidly, than natural radioactivity does? On Earth, uranium and thorium slowly break down into less massive nuclei and even, very occasionally, undergo spontaneous fission, but hydrogen remains stable and shows no signs whatever of fusing.

The reason for this is not difficult to see. Massive nuclei such as those of uranium and thorium have all of their protons and neutrons squeezed together into one place. Any changes that might take place among them *do* take place. In the case of hydrogen fusion, however, two hydrogen-2 nuclei, or four hydrogen-1 nuclei, which exist apart from each other, must possess enough energy to break through the electron barrier, overcome the mutual repulsion of the nuclei for each other, and then collide with sufficient force to initiate fusion. At ordinary temperatures, their motions simply don't involve more than the tiniest fraction of the required energy.

In order to supply the required energy, the temperature must go very high indeed—to the millions of degrees. Even then, it helps to compress the hydrogen to very high densities so that there will be enormous numbers of collisions as hydrogen nuclei jiggle back and forth across the abnormally small distances separating them.

These conditions are satisfied in the centers of stars. In 1926, Eddington produced a convincing line of argument to show that the Sun was gaseous throughout. At the center of the Sun, the temperatures and pressures were so high that the atoms broke down. The electrons were crushed together and the nuclei approached one another freely.

We now believe that the center of the Sun has a temperature of about 15,000,000° C. and a density of about 160 grams per cubic centimeter, or about 8 times the density of gold. (Yet that center is gaseous because the atoms are broken and the nuclei move about as freely as intact atoms do in ordinary gases.) As a result, nuclear fusion takes place at the surface of the small core of helium at the Sun's center—the helium that was part of the Sun originally plus an additional quantity formed by hydrogen fusion over the last 4.55 billion years.

If, then, we wanted to have a fusion reaction here on Earth, how could we manage it? How could we get a temperature and pressure high enough?

Once the fission bomb had been developed, it could be seen that this was one way of developing the necessary temperatures and pressures. If the fission bomb included a quantity of hydrogen in some form, the first few instants of the fission reaction might raise the hydrogen to the temperatures and pressures required to ignite a fusion process.

In 1952, both the United States and the Soviet Union developed a successful nuclear fusion bomb, which is more popularly known as a hydrogen bomb or an H-bomb. It is sometimes called a thermonuclear bomb, where thermo- is from the Greek word for "heat," in that the fusion bomb is ignited by extreme heat rather than by neutron bombardment.

The vast energies of the fusion bomb cause it to explode with enormous fury. The explosion is so powerful that in any war in which it is freely used, civilization will surely be destroyed almost at once; perhaps humanity in general will be, and, in the extreme, most or all of life.

The Sun itself is, in effect, a vast fusion bomb, but it does not blow apart. The Sun, which is 333,000 times as massive as the Earth, has an enormous gravitational field that holds it together against all of the fury of fusion energy in its interior. And so we sit and bask in the welcome light and warmth of this cosmic bomb—although it is well that we are at a safe distance of 93,000,000 miles from it.

Can nuclear fusion be ignited and kept going in a *controlled* fashion? Can it be made to develop energy slowly, energy that can be used and that will not be destructive? If this can be done in practical fashion, then we will have a form of nuclear energy in which the fuel is easy to get and easy to handle. Instead of having to obtain uranium and thorium from rocks in which it is thinly spread, we would get hydrogen-2 out of the ocean. (Hydrogen-2 is much rarer than the ordinary hydrogen-1, but hydrogen-2 will undergo fusion more easily, and there is enough of it, comparatively rare though it is, to last us for billions of years.)

Then, too, fission energy requires a rather sizeable minimum supply of fissionable material, creating the possibility of runaway reactions and meltdowns if not properly controlled. Fusion energy can be carried through with small quantities of fuel, which would make the possibility of a major accident a thing of the past. Finally, fusion energy would not produce radioactive materials with the profusion that fission energy does.

In order to produce controlled nuclear fusion, however, we must subject hydrogen-2 to high temperatures and pressures. We cannot yet manage the pressures, however, and we must raise the temperature now possible still higher, while at the same time keeping the hydrogen confined within a magnetic field.

For more than thirty years scientists have labored at

the task of producing controlled fusion, and have gotten ever closer—but they have not yet reached this goal.

Breakdown Particles

It was mentioned earlier that helium-4 is a particularly stable nucleus. Thus, soon after the Universe was formed, the four simplest nuclei were formed. Hydrogen-1 existed first because its nucleus was a mere proton. By adding a neutron, it became hydrogen-2 (1 proton, 1 neutron); which, by adding a proton, became helium-3 (2 protons, 1 neutron); which, by adding a neutron, became helium-4 (2 protons, 2 neutrons). Hydrogen-2 and helium-3, while stable nuclei, have fairly low packing fractions, which is why they had a considerable tendency under the conditions of the early Universe to change over to the stable helium-4. The result was that 90 percent of the atoms in the Universe today are hydrogen-1 and 9 percent are helium-4. Everything else makes up the remaining 1 percent.

Moreover, the early Universe never got past the helium-4 level. Helium-4 is so stable that it has virtually no tendency to add either a proton, a neutron, or another helium-4. The nuclei that would form—lithium-5 (3 protons, 2 neutrons), helium-5 (2 protons, 3 neutrons), or beryllium-8 (4 protons, 4 neutrons)—are all so unstable that they have half-lives of anywhere from a hundredth of a trillionth of a second to less than a billionth of a trillionth of a second. As a result, all nuclei with mass numbers beyond 4 have been formed (as I mentioned earlier) at the centers of stars, where conditions make such formation not only possible, but likely.

Among the higher mass numbers, nuclei that might be viewed as made up of helium-4 units are particularly stable.

194

Carbon-12 (6 protons, 6 neutrons—3 helium-4 units) hangs together tightly. So does oxygen-16 (8 protons, 8 neutrons—4 helium-4 units). Both have lower packing fractions than do their neighbors.

As nuclei grow more massive, this helium-4 effect diminishes. Nevertheless, nuclei such as neon-20 (5 helium-4 units), magnesium-24 (6), silicon-28 (7), sulfur-32 (8), and calcium-40 (10) are particularly stable. All of these nuclei, from helium-4 to calcium-40, are the most common isotopes of their elements.

Beyond calcium-40, however, the helium-4 unit seems to lose its stabilizing effect. Apparently, as the number of protons in the nucleus grows, it is no longer sufficient to have an equal number of neutrons in order to make the nucleus stable. There must be an excess of neutrons.

Thus, in iron-56, the most common iron isotope, there are 26 protons and 30 neutrons, which makes the neutron/proton ratio 1.15. In tin-118, the most common tin isotope, there are 50 protons and 68 neutrons, a neutron/proton ratio of 1.36. In gold-197, the only stable gold isotope, there are 79 protons and 118 neutrons for a neutron/proton ratio of 1.49. The most massive stable nucleus is bismuth-209, with 83 protons and 126 neutrons for a neutron/proton ratio of 1.52.

Beyond bismuth-209, no excess of neutrons will suffice to keep a nucleus stable. Thus, uranium-238 has 92 protons and 146 neutrons for a neutron/proton ratio of 1.59, but even that large excess does not suffice to keep the nucleus entirely stable.

The German-American physicist Maria Goeppert-Mayer (1906–1972) tackled the problem of why some stable nuclei are more stable than others. She suggested that there are nuclear shells and subshells, as there are electron shells and subshells. She worked out the numbers of protons and neutrons it took to fill these shells, and pointed out

195

that filled shells produce nuclei more stable than their neighbors.

The number required to fill a nuclear shell is a shell number, which is sometimes called a *magic number*. (This last term is inappropriate in that there is no "magic" in science—but scientists are as prone to be dramatic as other human beings.) The German physicist Johannes Hans Daniel Jensen (b. 1907) worked out the notion of shell numbers independently of Goeppert-Mayer, and the two shared a Nobel prize for this work in 1963.

When a massive nucleus is so massive that it is unstable, there is a natural tendency for it to lose particles in order to become a less massive, and therefore more stable, nucleus. An efficient way for it to do this is to get rid of an alpha particle (a helium-4 nucleus) as a breakdown particle. Such a nucleus clings together so tightly that it is easy to expel as a unit, and the mass number decreases by four at a stroke. Therefore, uranium-238, uranium-235, thorium-232, radium-226, and many other nuclei more massive than bismuth-209, emit alpha particles.

Nuclei less massive than bismuth-209 do not usually emit alpha particles. Neodymium-144 is about the lightest nucleus that does, but it emits very few because its half-life is about 2,000 trillion years.

Of course, radioactive atoms often emit beta particles in the course of breakdown, which raised a problem. In the 1920s, the existence of beta particles was considered excellent evidence that the nucleus contained electrons. After all, if a dime falls out of your change purse, it can only be because a dime was in your change purse to begin with.

However, such homey analogies don't always work, which is why "common sense" is so often a dangerous guide as applied to science. By 1932, scientists were convinced that the nucleus contained only protons and neutrons—no electrons. Where, then, did the beta particles come from? If there are no electrons in the nucleus, we can only as-

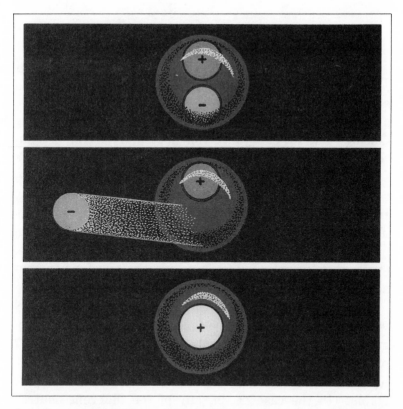

A neutron in its stable position inside a nucleus can be thought of as a neutral combination of a proton and an electron. Separated from the nucleus, the neutron breaks down into a proton and an electron, with a half-life of twelve minutes.

sume that one is created there and is immediately ejected. But how?

Suppose we consider the neutron an uncharged particle, not because it has no electric charge, but because it has both a positive and a negative charge that neutralize each other. If the negative charge is ejected in the form of an electron, then the positive charge is left, whereby the neutron has become a proton. (The situation is actually more complicated than that, as we shall see, but for now this viewpoint will do.)

Why should such a change take place, however? It doesn't take place in the various nuclei we find to be stable. These could remain unchanged for indefinite periods, even eternally, for all anyone could tell in the 1930s. Yet there are nuclei that eject electrons—some slowly, some quickly—and for every electron ejected, a neutron in a nucleus has changed to a proton. To answer the question, let's suppose that a nucleus of a given atomic number must contain a fixed number of protons equal to that atomic number. In addition, it must contain a certain number of neutrons if the nucleus is to be stable. Sometimes only one certain number of neutrons will do. Thus, in the case of fluorine, every atomic nucleus must have exactly 9 protons and exactly 10 neutrons if it is to be stable.

However, sometimes there is a certain flexibility where neutron numbers are concerned. Thus, every nitrogen nucleus must have 7 protons, but it can have either 7 or 8 neutrons and be stable. Every oxygen nucleus must have 8 protons, but can have 8, 9, or 10 neutrons and be stable. (In the case of tin, any of ten different neutron numbers will keep the nucleus stable.)

But what if a nucleus has too many neutrons to make stability possible? For instance, hydrogen-1 is stable, with its nucleus consisting of 1 proton alone. So is hydrogen-2, with a nucleus consisting of 1 proton and 1 neutron. Hydrogen-2 is less stable than hydrogen-1, but, nevertheless, stable—and if left to itself will exist unchanged indefinitely.

(How can one nucleus be less stable than another, yet stable? Imagine a coin resting in the center of a table. Left to itself it will remain there indefinitely. Another coin might be resting near the edge of the table. It, too, will remain there indefinitely—yet, while stable, it is less stable than the centered coin for it will take a smaller disturbance to knock it off the table. In the same way, hydrogen-2 can more easily be made to undergo fusion than hydrogen-1, which is why there is so much less hydrogen-2 in the Universe than hydrogen-1, even though both are stable.)

Hydrogen-3, with its nucleus consisting of 1 proton and 2 neutrons, is not a stable nucleus; there are too many neutrons for that. One might expect the hydrogen-3 nucleus to eject a neutron, but that would involve a high energy of activation—energy not available to the hydrogen-3 nucleus under ordinary conditions. A second alternative is for one of the neutrons of the hydrogen-3 nucleus to be converted to a proton, and for a beta particle to be ejected. This requires a rather small energy of activation; the half-life of hydrogen-3 is only about 12¼ years. After the ejection of the beta particle, the nucleus contains two protons and one neutron, having become the stable isotope helium-3.

Similarly, there are only two stable carbon isotopes: carbon-12 (6 protons, 6 neutrons) and carbon-13 (6 protons, 7 neutrons). In 1940, the Canadian-American biochemist Martin David Kamen (b. 1913) identified carbon-14 (6 protons, 8 neutrons), which has one neutron too many. The emission of a beta particle, converting one of the neutrons to a proton, however, produces stable nitrogen-14 (7 protons, 7 neutrons). There are many other examples of this type.

A neutron is more massive than a proton and an electron put together. Therefore, if a neutron is converted into a proton via an ejected electron, there is an overall loss of mass and a dissipation of energy. Will a free neutron undergo a spontaneous conversion to a proton, then, giving off electrons as it does so?

It was at first difficult to test this hypothesis because even if a stream of neutrons was produced, the neutrons generally collided with, and were absorbed by, other nuclei before they had a chance to break down. It was not until 1948 that the difficulty was overcome by a strong beam of neutrons passed through a large, evacuated cylindrical tank. There was an electric field around the tank so that any electrons that might be produced would curve off in one direction, while protons would curve off in the other. The proposed breakdown was then indeed observed, with the neutron breaking down into a proton and electron with a half-life in the neighborhood of 12 minutes. (This is not the full description of what happened, but it will do for now.)

If this is the case, why don't neutrons break down in every nucleus, until there is nothing left but protons? Apparently, neutrons within the nucleus are in close association with protons, and under such conditions, provided the number of neutrons is neither too many nor too few, the neutrons are stable. (There will be more to say about this later.)

Spontaneous breakdowns of isolated particles, when they take place, always seem to result in a reduction of mass—meaning that a neutron can break down to the less massive proton, but a proton cannot break down to the more massive neutron.

But in that case, why doesn't the proton break down to the less massive electron, releasing all but $1/1836$ of its mass as energy? The answer to this question is that there are conservation laws that seem always to be obeyed. There is, for instance, the conservation of electric charge, which states—if uncounted numbers of laboratory observations are to be trusted—that a positive electric charge, left to itself, can neither be created nor destroyed. The same is true for a negative electric charge.

In the 1930s, the only two charged particles known were the proton, which was positively charged, and the electron, which was negatively charged. (This is no longer the situation, but let us suppose it is and answer the question on that basis. We can always qualify the answer later.) If the only way the proton can lose mass is to be converted to an electron, then the positive charge of the proton must be destroyed and the negative charge of the electron must be created. Neither is possible. For this reason, the proton cannot break down into any other particle, for there is no less massive particle that contains a positive charge (so it was thought).

In the same way, an electron cannot break down into smaller particles, in that the only smaller particles known in the 1930s were the photon and the graviton, both of which have zero mass and zero charge. The electron, in breaking down into either of these, must have its negative charge destroyed—and this is impossible. For this reason, an electron cannot break down.

Notice that when a neutron breaks down into a proton (whether we are talking of a free neutron or of one that is part of a nucleus), an electron is formed at the same time. In this way, the uncharged neutron (0) forms a positive charge ($+1$) and a negative charge (-1). The two, taken together, are still zero: $0 = (+1) + (-1)$. (The law of conservation of charge will allow the production or destruction of pairs of opposite charge, but will not allow the production or destruction of one without the other.)

We might ask, then, why a photon can't turn into a graviton, or vice versa. In the case of these two particles, there is no electric charge to worry about. There is, however, spin, or angular momentum. The law of conservation of angular momentum tells us that spin can be neither created nor destroyed. A photon with a spin of 1 can't break down into a graviton with a spin of 2, or vice versa. There might be other factors that would prevent the change, but the matter of spin is, in itself, sufficient.

Thus, in the early 1930s there were just five known particles of which the Universe might be composed. Four of these, the proton, electron, photon, and graviton, are stable. The fifth, the neutron, is unstable. This view was not to endure for long.

8

ANTIMATTER

Antiparticles

The list of particles making up the universe—proton, neutron, electron, photon, and graviton—seems peculiar in one way. Why should the positive electric charge be housed in protons and the negative electric charge in electrons when protons are 1,836 times as massive as electrons?

The two particles, so different in mass, have electric charges that are precisely equal in size, even if they are opposite in nature. We can tell this because the hydrogen atom, made up of 1 proton in its nucleus and 1 electron outside the nucleus, is exactly neutral. No excess electric charge, either negative or positive, has ever been detected to even the tiniest degree.

Nor have scientists found any basic difference in the two types of charge that make it necessary for the electron to be associated with very little mass, while the proton is associated with much greater mass. In short, the proton and electron make an unlikely and puzzling pair.

The puzzle began to come together in the late 1920s, when Dirac attempted to study the properties of the electron by working out the mathematics of its wave properties. It seemed to him that it should be possible for the electron to exist in one of two energy states, one the opposite of the other. Naturally, Dirac's first thought was that the electron itself represented one of the energy states, and the proton the other. This would be delightful, if true, for it would simplify the Universe still further, in that the electron and proton would then be simply two different states of one fundamental particle.

That was too good to be true, however, for Dirac quickly saw that the equations would not be truly satisfied unless the two states were exactly similar in every way but for some crucial orientation. They would have to be mirror images of each other, so to speak, like your two hands, which are in overall form exactly alike in every way, except that one is thumb-rightward and the other is thumb-leftward.

If electric charge were subject to mirror image variation in the electron—positive in one state and negative in the other—that was all that could vary. Everything else had to be identical. Not only would the size of the charge in the two states have to be identical, but so, too, would its mass. In 1930, then, Dirac suggested that there must exist a particle exactly like the electron in every way, except that it carried a positive electric charge of precisely the same size as the electron's negative electric charge.

The same sort of argument would lead to the conclusion that there must exist a particle exactly like the proton in

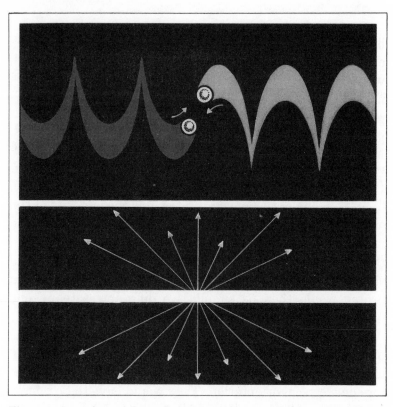

The meeting of particle and antiparticle is something like having two opposite waves cancel each other out. The total charge of the two particles is +1 + −1, or 0. In the course of mutual annihilation their mass is completely converted into energy in the form of gamma rays.

every way, except that it carried a negative electric charge of precisely the same size as the proton's positive charge.

In general, a particle that is exactly like another except for being opposite in one key respect has come to be called an antiparticle, where the prefix anti- is from a Greek word for "opposite." The positively charged electron would be an antielectron, and the negatively charged proton would be an antiproton.

If a particle and an antiparticle meet, it is something

like having two opposite waves (one going up, where the other goes down, and vice versa) meeting. Just as two waves can cancel each other out into a straight line going neither up nor down, so that there is no wave at all, the particle and antiparticle cancel each other out, leaving no particle at all. This is called mutual annihilation.

Interestingly, this phenomenon does not violate the law of conservation of electric charge, for when a particle and antiparticle meet, the total charge of the two particles is $(+1) + (-1)$, or 0. Once they undergo mutual annihilation, the electric charge that remains is still 0; therefore, the conservation law has not been violated. It is only the positive charge *alone* or the negative charge *alone* that can be neither created nor destroyed. Positive charge *and* negative charge *together* can be created or destroyed in any quantity.

It is only the electric charge that disappears in mutual annihilation, of course, for that is the only part of the particle and antiparticle that is of opposite character. Both the particle and antiparticle have identical mass, and this double mass cannot disappear. Mass, however, is a form of energy, and it can change its form. In the course of mutual annihilation, then, gamma rays appear with an energy exactly equivalent to the mass that existed before the annihilation.

The opposite change can also take place. If enough energy can be concentrated into a tiny spot, it can be converted into matter. When this takes place, however, a particle alone or an antiparticle alone cannot be formed, for in either case, electric charge would have been created out of nothing. One can only produce a particle *and* an antiparticle *together*, so that the total electric charge remains zero. This is called pair production.

Dirac's theory was an extremely interesting one, mathematically, yet mathematics, however interesting, does not carry much weight if it can't be matched with reality. For instance, scientists are certain, for a variety of mathematical and theoretical reasons, that gravitons (or gravitational waves, depending on whether you look at them as particles or as waves) must exist. To be sure, theory also tells us that gravitational waves are so lacking in energy that detecting them might be next door to impossible. (It would be rather like trying to pick up a single dust particle with a monkey wrench. You can't do the job until you devise a sufficiently delicate pair of tweezers.)

That is exactly what scientists are trying to do. Despite their certainty as to the existence of gravitational waves, there are those who bend their efforts toward the construction of devices that will actually detect these waves. For years they have failed, but scientists are confident that someday they will succeed. When success comes, it will crown theory with observation, and this will mean great jubilation and a Nobel prize for the detectors.

So it was with Dirac's suggestions concerning antiparticles. In the world about us there were electrons in countless numbers, but no antielectrons had ever been observed at the time Dirac's theories were announced. Until the antielectron was actually detected, the work could not be taken entirely seriously. However, the antielectron was soon detected, under circumstances that make it appear as though it would have been detected even without Dirac's published results. To see how this was done, we must again backtrack.

Cosmic Rays

A charged electroscope with its gold leaves spread wide apart tends to lose its charge slowly, even when there are no radioactive materials in the near vicinity. This was not a great surprise to early investigators because it was quite likely that small quantities of radioactive material were widespread in common soil. Even if they were present in quantities too small to detect in ordinary ways, the occasional speeding particle would drain off a bit of the electroscope charge, and eventually the electroscope would be entirely discharged.

But then investigators found that there seemed no way of stopping the discharge. If the electroscope were taken over stretches of water well away from land, the charge still disappeared slowly. If the electroscope were encased in a thickness of lead sufficient to block the passage of radiation, the charge still disappeared, although more slowly than before.

The Austrian physicist Victor Franz Hess (1883–1964) thought he would investigate the subject by sending an electroscope high into the atmosphere in a hot-air balloon. In that the radiation sources were assumed to be almost entirely in soil, moving far up above the soil ought to stop the loss of charge more efficiently than anything else tried hitherto.

In 1911, Hess made the first of ten balloon ascensions, taking the electroscopes up to six miles above Earth's surface. To his astonishment, Hess found that the electroscopes lost charge *faster* the higher he went. Hess could see no way of explaining this except to suppose that a very penetrating radiation must be coming from outer space. For this discovery, Hess received a share of a Nobel prize in 1936.

In 1925, the American physicist Robert Andrew Millikan (1868–1953) interested himself in this penetrating radiation from the sky, to which he gave the name cosmic rays because they seemed to originate from somewhere in the cosmos. It seemed quite certain to Millikan that cosmic rays were a form of electromagnetic radiation even shorter

When charged, the gold leaves of an electroscope spread wide apart in mutual repulsion. The closer they are to a radioactive source, the quicker they lose their charge and collapse. In 1911, Victor Franz Hess took one up in a hot-air balloon. He was attempting to remove

in wavelength, and therefore higher in energy and more penetrating, than gamma rays.

On the other hand, Compton rather suspected that cosmic rays were streams of electrically charged subatomic particles that, if moving rapidly enough, could also be more penetrating than gamma rays. (This resembled the argument, a generation before, over whether cathode rays were waves or speeding particles.)

it from the natural radioactivity of common soil. He actually found that it lost its charge faster the higher it went. In the illustration, the vertical lines represent cosmic rays.

How could the dispute be settled? Well, if cosmic rays were a form of electromagnetic radiation, as Millikan thought, they would be electrically uncharged and would be unaffected by Earth's magnetic field. If they came from all parts of the sky in equal quantity, they would reach all parts of Earth's surface in equal quantity. If, however, cosmic rays were charged particles, they would be deflected by the Earth's magnetic field and would reach Earth in greater quantities the farther you went from the equator and the closer you came to the magnetic poles. In other words, the higher the latitude, the greater the concentration of cosmic rays. This was called the latitude effect.

Compton traveled over the world measuring cosmic ray incidence here and there to see if the latitude effect really existed. By the early 1930s, he was able to show that it *did* exist and that the cosmic rays *were* electrically charged particles.

In 1930, the Italian physicist Bruno Benedetto Rossi (b. 1905) pointed out that if cosmic rays were positively charged, more of them would come from the west than from the east. If they were negatively charged, the reverse would be the case. In 1935, the American physicist Thomas Hope Johnson (b. 1899) was able to show that more came from the west, and that cosmic-ray particles carried a positive electric charge.

We now know that cosmic ray particles are speeding atomic nuclei issuing from stars. In that stars are composed mostly of hydrogen, cosmic-ray particles are mostly hydrogen nuclei; that is, protons. They also contain some helium nuclei and a thin scattering of more massive nuclei.

Our Sun emits a constant stream of speedy protons and other charged particles, as Rossi demonstrated in the 1950s. This is now called the solar wind. Particularly violent disturbances on the solar surface, such as solar flares, produce a shower of such particles with greater energy than usual. The greater the energy, the greater the velocity, and when these velocities approach the speed of light they are classified as cosmic-ray particles. The Sun occasionally emits particles that just barely fall into this classification.

Hotter and more violent stars than our Sun are more copious emitters of cosmic-ray particles; supernova explosions are particularly good sources. Once cosmic-ray particles are speeding through space, they can be accelerated

210

and made even more energetic by the magnetic fields of the stars they pass, as well as by the overall magnetic field of the Galaxy.

As a result, cosmic ray particles are more energetic than the radiations obtained from radioactive materials. This offered nuclear physicists a new and more powerful tool, for cosmic ray particles can bring about nuclear reactions that radioactive radiations are not energetic enough to start.

To balance this, however, cosmic ray particles are not as easily handled as radioactive materials, which can be concentrated and worked with in the laboratory; radioactive radiations can be called on at will and carefully aimed. Cosmic ray particles, however, come at their own time and can only be dealt with in more concentrated form by climbing mountains or going up in a balloon.

The American physicist Carl David Anderson (b. 1905), a student of Millikan, also studied cosmic rays. He allowed them to pass through a cloud chamber and hoped that by following the curvature of the lines of fog droplets they produced he could learn something about them. The cosmic rays were so energetic, however, that they passed through the cloud chamber too quickly to have time to curve appreciably in response to the magnetic field. Anderson therefore devised a cloud chamber with a lead barrier running across the center. Cosmic rays striking the lead would be energetic enough to smash through, but they would lose enough energy in the process to curve markedly thereafter in response to the magnetic field.

In 1932, Anderson noted, emerging from the lead barrier, a curved pathway that looked precisely as though it had been caused by a speeding electron. However, it curved in the wrong direction! Anderson realized that he was observing the path of a particle just like an electron, but that carried a positive charge. It was the antielectron that Dirac

211

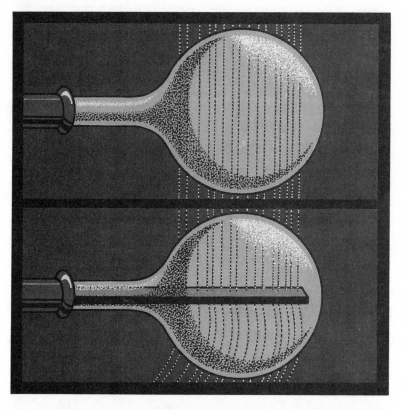

Cosmic rays pass through a cloud chamber so quickly that their path is not significantly changed by a magnetic field. Carl David Anderson placed a lead barrier inside the chamber, slowing them down enough to be studied.

had suggested, in theory, two years before. As a result, Anderson shared in a Nobel prize with Hess in 1936.

The particle Anderson found was referred to as a positive electron, or positron. This, in my opinion, is a faulty word formation, as well as a poor choice. For one thing, the common ending for subatomic particles is -on. We have electron, neutron, proton, photon, and graviton as examples. The *r* in electron and neutron belongs to the root of the word, as in *electricity* and *neutral*. For this reason, if the positive electron were to be given a subatomic name, it should be called a positon, without the *r*, because *positive* contains no *r*. Moreover, whether positon or positron, such a name obscures the relationship to the electron. The particle should be called an antielectron, for all other antiparticles, without exception, add the prefix anti- to the name of the particle that is their opposite. Nevertheless, the name positron has been used so commonly and so often that there is no longer any hope of changing it.

(It often happens that a poor name is given to an object or a phenomenon to begin with, either out of ignorance or out of bad judgment. Sometimes, it can be changed in time, but often the ill-chosen name is used so commonly by so many that it becomes inconvenient or even impossible to change it.)

The positron behaves exactly as Dirac's theory suggested. It quickly undergoes mutual annihilation when it encounters one of the numerous electrons in its immediate environment, producing gamma rays of energy exactly equal to that of the combined mass of the electron and proton. It was quickly found, too, that if alpha particles were allowed to strike lead, some of the particle energy could be converted into an electron-positron pair that would emerge from the lead, with tracks curving in opposite directions. This takes us back to the question we had raised earlier. What happens if we have a radioactive isotope that contains too few neutrons for stability?

The easiest way of producing an additional neutron inside a nucleus is to convert one of its protons into a neutron. This yields a nucleus with one additional neutron and one fewer proton, which is what might be called for for stability.

For example, phosphorus-30 has 15 protons and 15 neutrons in its nucleus. The only stable phosphorus isotope is phosphorus-31, with 15 protons and 16 neutrons in its nucleus. In other words, phosphorus-30 has too few neutrons for stability. Suppose, however, that one of the protons in the phosphorus-30 nucleus is converted into a neutron. The positive charge on the proton cannot be destroyed (in line with the law of conservation of electric charge), so it must show up elsewhere. But what if the nucleus emits a positron, a type of positive beta particle? This takes care of getting rid of the positive charge. If phosphorus-30 emits a positron, then instead of having 15 protons and 15 neutrons in its nucleus, it has 14 protons and 16 neutrons, which is the stable silicon-30.

Thus, in 1934, when the Joliot-Curies discovered artificial radioactivity in the form of phosphorus-30, they were also producing a new type of radiation that turned out to be a stream of speeding positrons. Here was a way of producing positrons without having to bombard material with either cosmic ray particles or alpha particles. They had formed a neutron-deficient nucleus and had allowed it to undergo radioactive decay.

Positron emission by a nucleus accomplishes results that are just the opposite of electron emission. Whereas electron emission causes the atomic number to go up by one as an additional proton is formed from a neutron, positron emission causes the atomic number to go down by one as a proton is lost through conversion into a neutron.

There might seem to be a puzzle here. Because the neutron is slightly more massive than the proton I have emphasized that the neutron spontaneously decays into a proton, but that a proton does not decay, "uphill," into a neutron.

This, however, is true only if we are speaking of free particles. In a nucleus, where protons and neutrons exist in association, what counts is the mass of the entire nucleus. In a neutron-deficient nucleus, the total mass of the nucleus can go down if a proton changes into a neutron because the packing fraction increases. Therefore, the change can occur.

That is what makes particular isotopes unstable. If the total mass of a nucleus is going to decrease if a proton changes into a neutron, or if a neutron changes into a proton, the appropriate change will take place. If an isotope has a mass that will be increased if either a neutron changes into a proton or a proton into a neutron, then neither change will take place and the isotope will remain stable.

As it happens, when there are either 43 or 61 protons in a nucleus, no matter how many neutrons there are present, a neutron-proton change, one way or the other, will always lower the total mass. This is why there are no stable isotopes of technetium (43) or promethium (61).

Another way in which a proton within a nucleus can be changed into a neutron is for the nucleus to pick up one of the electrons outside the nucleus, thus neutralizing the electric charge of one of its protons and converting it into a neutron. Almost always, the electron is picked up from the K shell, which is the electron shell nearest the nucleus. The process is therefore known as K capture, which was first observed in 1938 by the American physicist Luis Walter Alvarez (1911–1988). This is much less likely to happen than positron emission, however.

There is no theoretical reason why the reverse of proton-to-neutron conversion wouldn't be possible. To convert a neutron into a proton, a nucleus, instead of emitting an electron, might pick up a nearby positron. The only trouble here is that in the ordinary matter about us there are no positrons to speak of, so the chance of positron capture is nil.

Particle Accelerators

Once the antielectron, or positron, was produced and observed, scientists could feel entirely confident that an antiproton must also exist. Confidence, however, is not enough. They wanted to observe one in existence.

However, antiprotons don't seem to exist about us any more than do antielectrons. They must be formed in some type of nuclear reaction, and then observed, but this is easier to say than to do. A proton is 1,836 times as massive as an electron; we can, therefore, be reasonably certain that an antiproton must be 1,836 times as massive as an antielectron.

Scientists could form electron-antielectron pairs by smashing alpha particles into lead, but in order to form a proton-antiproton pair, one must find projectiles with 1,836 times the energy of the alpha particles that suffice for the smaller task. Such energetic alpha particles, unfortunately, do not exist.

There undoubtedly exist cosmic ray particles sufficiently energetic to do the job, but they are a great deal fewer in number than those energetic enough to bring about the production of a positron. It would be a long wait for one of the rare particles to happen to come along and form an antiproton exactly where it could be detected.

In the late 1920s, however, physicists began to work at the task of creating energetic projectiles of their own. For the purpose, you would want a massive particle to begin with, in that the energy of a speeding particle increases with its mass. That means something at least as massive as a proton, which is a natural choice because all one need do to obtain a supply of protons is to remove the outer electron from hydrogen atoms, something that can be done without much trouble. Alpha particles would be more massive still, of course, but they be from the nuclei of helium, which is a substance much rarer than hydrogen and much more difficult to strip down to its bare nucleus.

Given a supply of protons, one would then pass them through a magnetic field to accelerate them and make them move faster. The stronger the magnetic field, the more sharply a proton is accelerated. If the accelerated proton is allowed to smash into an atomic nucleus, a nuclear reaction is possible. A device that accomplished this was called an atom smasher by the newspapers in the early days of this work, but the term is overly dramatic. The proper name is the soberly descriptive particle accelerator.

The first useful particle accelerator was devised in 1929 by the British physicist John Douglas Cockcroft (1897–1967) and his co-worker, the Irish physicist Ernest Thomas Sinton Walton (b. 1903). Using their particle accelerator, they bombarded nuclei of lithium-7 (3 protons plus 4 neutrons) with energetic protons. In the process, a proton would smash into a nucleus and remain there, forming beryllium-8 (4 protons plus 4 neutrons). Beryllium-8 is, however, extremely unstable, and in about a billionth of a trillionth of a second splits up into two helium-4 nuclei (2 protons plus 2 neutrons). This was the first nuclear reaction brought about by an accelerated particle, for which Cockcroft and Walton shared a Nobel prize in 1931.

In the years immediately following this feat, other types of particle accelerators were developed. The one that yielded the most fruitful results was devised in 1930 by the American physicist Ernest Orlando Lawrence (1901–1958). He tackled the problem of an ordinary magnetic field causing a proton to accelerate forward in a straight line, quickly passing beyond the field and accelerating no farther. The field would have to be extended a great distance in order for the proton to continue to accelerate.

Lawrence devised a way of flipping a field back and forth so that it forced a proton to follow a curved path one way, and then to follow another curved path the opposite way, completing one "cycle" and staying within the magnetic field throughout. By repeating this over and over, the particle would travel in slowly expanding cycles. Although the particles would have to cover a greater and greater distance as the cycle expanded, they would go just enough faster to complete the circuit in the same length of time, staying in step with the field as it flipped back and forth. The particles would remain within the field for a considerable length of time, then, even though the device itself wasn't very large. In this way, a small device could produce particles of unexpectedly high energies. Lawrence called his device a cyclotron, receiving a Nobel prize in 1939 for its development.

Larger and more powerful cyclotrons were quickly built. New designs were used in which the magnetic field grew more powerful as the particles speeded up. This kept them moving in tight cycles so that they didn't expand out of the field until scientists were ready to have them do so. These proton synchrotrons created even more energetic particles. It was then possible to have two cycling sets of particles moving in opposite directions and eventually colliding head on. This doubled the former energy production possible through a single stream of particles hitting a stationary object.

In 1987, the United States began to consider spending some six billion dollars for the construction of a superconducting super collider, a particle accelerator in which particles will be sent around a track 52 miles long, producing ten times as much energy as any other particle accelerator in existence. (Later in the book there will be occasion to mention what it is hoped will come of such an unimaginably powerful device. Of course, there are very occasional cosmic ray particles with several million times the energy that even this accelerator will produce, but waiting for such particles to turn up would be a very long and thankless job.)

As long ago as the early 1950s, particle accelerators had been built that could produce particles energetic enough to form a proton-antiproton pair. Naturally, if they struck an appropriate target, such energetic particles would produce all types of nuclear reactions and result in the formation of all types of particles. This mass of various particles could then be allowed to stream through a magnetic field. All of the positively charged particles would curve in one direction, and all of the negatively charged ones in another. The most massive negatively charged particles were expected to be antiprotons; they would curve the least. At a considerable distance from the target, all of the particles would curve away and out of the field, with only the antiprotons (if any had formed) remaining.

In 1955, Segrè, the discoverer of technetium and now an American citizen, along with the American physicist Owen Chamberlain (b. 1920), did locate antiprotons in this manner. They shared a Nobel prize in 1959 for this work.

Baryons

Despite the fact that antielectrons (positrons), once formed, disappear almost at once through mutual annihilation with the first electron encountered, it is fair to consider an antielectron a stable particle. After all, an antielectron, *left to itself*, will remain an antielectron forever, as far as we know. It will never, of its own accord, change into anything else.

The reason for this is that any spontaneous change in a subatomic particle involves a loss of mass and the conversion of that mass into energy. The electron, however, remains, nearly a century after its discovery, the least massive object known that can carry a negative electric charge, while the antielectron is the least massive object known that can carry a positive electric charge. The only objects less massive than an electron or an antielectron are particles that, as far as we know today, have no mass at all—and none of them carries an electric charge. Therefore, the law of conservation of electric charge prevents electrons or anti-electrons from undergoing any spontaneous change, although one can annihilate the other by canceling out its respective charges.

But now the problem arises as to why the proton is stable. As long as the antielectron was not known, one might have maintained that the proton was the least massive object capable of carrying a positive electric charge so that it had to be stable by the law of conservation of electric charge. After 1932, however, this was no longer a valid argument. Why couldn't the proton decay into an antielectron? The positive electric charge would still exist, and almost all of the proton's mass would be converted to energy. Similarly, why couldn't the antiproton decay into an electron? The answer is that they just don't (and a lucky thing, too, or the Universe as at present constituted couldn't possibly exist, and neither could we).

From the fact that these particles don't decay in this manner, and from data gathered concerning all of the nuclear reactions that scientists have studied in all of the decades since subatomic particles began to be studied, it seems fair to conclude that what keeps protons stable is a conservation law. If it is not the conservation of electric charge, then it is the conservation of something else.

Electrons and positrons are examples of leptons (from a Greek word meaning "small"); they have small masses. Protons, neutrons, and antiprotons are baryons (from a Greek word meaning "heavy"); they are much more massive than leptons. The law of conservation that keeps a proton stable is the law of conservation of baryon number, which works as follows. A proton and a neutron are each given a baryon number of $+1$, and an antiproton a baryon number of -1. In that the electron and the antielectron are not baryons, they are each given a baryon number of 0. A proton and an antiproton, taken together, have a total baryon number of $(+1) + (-1) = 0$. Hence, the two can undergo mutual annihilation, leaving behind no proton and no antiproton for a total baryon number of 0, which is what they started with. The law of conservation of baryon number is not broken. In the same way, a large quantity of energy can create a proton-antiproton pair—baryon number 0 before and baryon number 0 after. Again, the law of conservation of baryon number is not violated. (In both cases, the law of conservation of electric charge is not violated either. Obeying one conservation law doesn't mean you don't have to obey the others. *All* conservation laws have to be obeyed where they are applicable. We shall see cases, however, in which a conservation law might *not* be applicable.)

A neutron (baryon number $+1$) can break down into the slightly less massive proton (baryon number $+1$) and an electron (baryon number 0), without breaking the law of conservation of baryon number or the law of conservation of electric charge.

A single proton (baryon number $+1$), however, cannot break down into a positron (baryon number 0) without breaking the law of conservation of baryon number; such a breakdown appears not to be possible. The law of conservation of electric charge is not broken, but that isn't enough. For the same reason, it would seem, a single antiproton (baryon number -1) does not break down into an electron (baryon number 0).

You might, of course, ask *why* there is such a conservation law. At the moment, scientists cannot answer this question. All they can say is that observation of nuclear reactions demonstrates that such a conservation law exists. (In recent years, however, there has been some question as to whether this conservation law is absolute or whether, under certain conditions, it can be broken.)

The proton is stable because it is the baryon with the

smallest possible mass, incapable of breaking down without losing its baryon status. The antiproton is stable because it is the antibaryon with the smallest possible mass. Of course, whenever you have a conservation law, you are bound to keep your eyes open for any apparent violation that might reveal some new fact or necessary modification of the law rendering a fairer (meaning "more just" and "more beautiful") view of the Universe.

For instance, a possible violation of the law of conservation of baryon number was uncovered in 1956 when a group of physicists discovered that when a proton and antiproton skimmed by each other closely without actually colliding, it was possible for the electric charges to cancel each other, while leaving the rest of the particles apparently untouched.

Without their electric charges, both the proton and the antiproton become neutral; thus it might be supposed that each has been converted into a neutron. But this can't be so. The proton and antiproton in combination have a baryon number of 0, while two neutrons have a total baryon number of $+2$. How can this be?

The answer is to suppose that when the electric charges cancel, the proton becomes a neutron (baryon number $+1$), whereas the antiproton becomes an antineutron (baryon number -1). The proton-antiproton pair (baryon number 0) becomes a neutron-antineutron pair (baryon number 0), with the law of conservation of baryon number upheld.

But how can there be an antineutron? The antielectron has an electric charge opposite to that of the electron, and the antiproton has an electric charge opposite to that of the proton. But neither neutron nor antineutron has an electric charge. What is it, then, that distinguishes the two?

Both neutron and antineutron, although they have no overall electric charge, have small local charges here and there that balance out over the entire particle (as we shall see). Both neutron and antineutron spin, and the effect of the spin on the small local charges is to create a small magnetic field. The neutron has its north magnetic pole pointing in one direction, and the antineutron, spinning in the same direction as the neutron, has its north magnetic pole pointing in the other direction. It is the direction of the magnetic fields that is opposite, not the spin, and it is this that differentiates the neutron and the antineutron.

The proton, neutron, and electron form the atoms, the planets, the stars—all of the matter we know. If the antiproton, antineutron, and antielectron existed in quantity in a place of their own, they would undoubtedly fulfill all of the functions of protons, neutrons, and electrons. The antiparticles could form antiatoms, antiplanets, antistars and, in general, antimatter.

This is not entirely theory. There are some (admittedly very simple) observations that support this view. In 1965, an antiproton and an antineutron were combined to form an antihydrogen-2 nucleus. Later on, two antiprotons and an antineutron were combined to form an antihelium-3 nucleus. (Each was actually simply an antinucleus.)

If we consider the various conservation laws, we know that however the matter of the Universe came into being, an equal quantity of antimatter must have been formed at the same time. If so, where is it?

We certainly know that Earth consists entirely of matter. In fact, we can be certain that the entire solar system, and even our galaxy, is formed entirely of matter, with antimatter existing, if at all, only in insignificant traces. If this were not so, there would be interaction between matter and antimatter now and then, and the constant formation of gamma rays. We don't, in fact, detect gamma rays reaching us from outer space in anything like the quantities we would expect if the galaxy contained substantial amounts of antimatter.

Some scientists have suggested that "original" matter and antimatter were separated, somehow, after formation, whereby there are even now clusters of galaxies made of matter and clusters of antigalaxies made of antimatter. They remain separate, therefore producing no significant level of gamma rays. Yet even this view seems unlikely because, if it were so, there should be a significant quantity of antiprotons, and antinuclei generally, in cosmic ray particles, some of which are sure to come from other galaxies— and there is not.

The reality might even be that two Universes were originally formed, one of matter and one (an "antiuniverse") of antimatter, with no communication between them possible. Undoubtedly, in such a case, the intelligent inhabitants of the antiuniverse, if there were any, would consider their own Universe to consist of matter and ours to be of antimatter. They would have as much right to maintain this as we have to maintain the reverse.

In recent years, however, as we shall see, new views on this subject have come into prominence. Scientists are willing to consider the possibility that matter and antimatter might *not* have formed in equal quantities to begin with.

9

NEUTRINOS

Saving the Laws of Conservation

All of the laws of conservation are important signposts in the understanding of nuclear reactions and of the behavior of subatomic particles. Anything that defies a law of conservation ought not to happen, which leaves only a limited number of possibilities of action. In other words, the laws of conservation prevent total anarchy and tell scientists, in effect, what they should look for.

Anything that *seems* to violate a law of conservation is therefore very unsettling. This is especially true if the law is the one considered most basic, most important, and, therefore, most inviolate. This is the law of conservation of energy, which, in the 1920s, seemed to be shaken.

In general, subatomic particles behave strictly in accordance with this law. If an electron and a positron annihilate each other, the energy produced, in the form of gamma rays, is exactly equal to the energy that had been present in the form of the mass of these two particles plus their kinetic energy as they approached each other. The same is true when a proton and antiproton annihilate each other.

Again, when a nucleus undergoes radioactive breakdown and emits an alpha particle, the new nucleus plus the alpha particle have a total mass slightly lower than the original nucleus. (This is why the nucleus originally breaks down, spontaneously, in this fashion.) The decrease in mass makes its appearance as the kinetic energy of the emitted alpha particle.

This means that all of the nuclei of a particular isotope that break down by alpha-particle emission give off alpha particles all traveling at the same velocity, and all equally energetic and penetrating. Measurements have shown that alpha-particle production always conserves energy.

Occasionally, a nucleus undergoing radioactive change produces two or more batches of alpha particles traveling at different speeds and possessing different energies. This implies that in such a situation the nucleus could exist at one of two or more energy levels. The one at the higher levels would produce alpha particles going at greater velocities than the ones at the lower energy level.

The situation is quite different where beta particles are concerned. When a nucleus gives off a beta particle (that is, a speeding electron), the new nucleus plus the beta particle is slightly less massive than the old nucleus. The difference in mass should be accounted for by the kinetic energy of the beta particle.

Sometimes the beta particle does move so quickly that it possesses a kinetic energy that just about balances the loss of mass. A beta particle never moves faster than this; it never produces kinetic energy greater than the loss in mass. If it did so, it would be creating energy out of nothing and would be violating the law of conservation of energy.

However, the beta particle usually moves more slowly, even much more slowly, than it should, and has less kinetic energy than is needed to balance the loss in mass. This violates the law of conservation of energy; too little is as bad as too much. When nuclei (all of the same sort) break down by emitting beta particles, they do so within a range of resultant particle speeds and kinetic energy. On average, the kinetic energy of the beta particles is only about one-third the quantity represented by the loss in mass. Energy, it seemed to those first studying the phenomenon, simply wasn't conserved.

For twenty years this range of velocity in the case of beta rays was observed and studied, remaining an absolute puzzle. Niels Bohr was so anxious about it he suggested that the law of conservation of energy might as well be abandoned, at least where beta particles were concerned. Few physicists, however, were ready to do that. (A general rule that works under all conditions but one should not be thrown out until every effort has been made to explain the exception.)

Pauli came up with, in 1930, an explanatory theory in regard to the beta emission conservation of energy problem. He suggested that whenever a nucleus gave off a beta particle, it gave off a second particle that carried off whatever quantity of energy the electron didn't. The kinetic energy of *both* particles, taken together, would then exactly account for the loss of mass in beta-particle production. The only problem with Pauli's notion, however, was this: If a second particle is produced, why is it never detected? The answer was that the electron carried off all of the charge required in the conversion of a neutron into a proton; the second particle, therefore, would have to be electrically neutral, which is a type of particle much more difficult to detect than an electrically charged one.

The neutron was detected—a neutral, massive particle capable of knocking protons out of nuclei. It was this capability that helped researchers locate the particle. In the case of this newly suggested particle in beta production, the small amount of energy it carries off is just enough to account for its speed; so that it could not have more than the merest trace of mass. In fact, some beta particles are given off at just about sufficient speed to account for *all* of the loss in energy to the nucleus, so that the second particle might have no mass at all.

A particle with no electric charge and no mass might seem to be difficult to detect, but this is not borne out by the example of the photon, which has neither electric charge nor mass and is very easy to detect. But then, the photon is a type of fuzzy wave packet that easily interacts with any bit of matter it encounters. What if the beta particle's companion was a minute particle that did not interact with matter?

In 1934, Fermi took up the subject of this particle, giving it the name neutrino (from the Italian word for "little neutron"). Fermi worked out in considerable detail what the particle's properties ought to be. He believed the particle to indeed be something with no mass, no electric charge, and virtually no tendency to interact with matter. It was a "nothing particle," or, as it was sometimes referred to, a ghost particle. It might as well not have been there, except that it served to balance the law of conservation of energy—a view that was not, in itself, very impressive. It could be argued (and some did argue in this fashion) that the position was invented just to save appearances. If the only way to save the law of conservation of energy was to invent a ghost particle, the law wasn't worth saving. However, the neutrino saved other conservation laws.

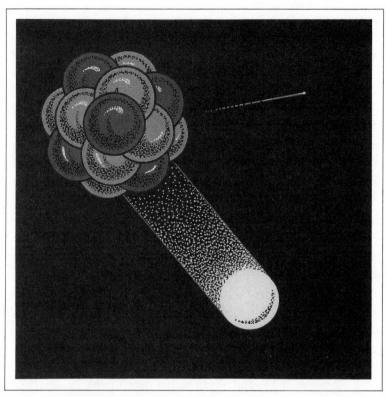

When a nucleus gives off a beta particle, the new nucleus plus the beta particle is slightly less massive than the old nucleus. The difference in mass, not completely accounted for by the kinetic energy of the beta particle, is assigned to the almost undetectable "ghost particle," the neutrino.

Consider, for example, a motionless neutron. Its velocity is zero; therefore, its momentum (which is equal to its mass times its velocity) is zero. There is a law of conservation of momentum that scientists understood even before the law of conservation of energy was grasped. In other words, whatever happens to a motionless neutron, the total momentum of the particles it gives rise to must remain zero—provided the rest of the universe doesn't interfere in any way.

After a certain time, the motionless neutron will break

down into a proton and an electron. The electron will go flying off in some direction at great speed and will, therefore, have a sizeable momentum. The neutron, now changed into a proton, will recoil in the opposite direction at a much slower speed, but will have a much greater mass. Ideally, the momentum of the electron (small mass × high velocity) should be equal to the momentum of the proton (large mass × low velocity). If the two dart off in exactly opposite directions, one has a momentum of $+x$ and the other of $-x$. These two momenta add up to 0; therefore, the law of conservation of momentum should be conserved.

But this is *not* what happens. The momentum of the electron is usually too low, and it and the proton do not move in precisely opposite directions. There is a bit of momentum not accounted for. If, however, we allow the existence of a neutrino, it might be moving in such a direction as to not only account for the missing energy, but for the missing momentum.

Then again, the neutron has a spin of either $+\frac{1}{2}$ or $-\frac{1}{2}$. Suppose it breaks down into a proton and an electron, and nothing more. The proton has a spin of $+\frac{1}{2}$ or $-\frac{1}{2}$, as does the electron. The total spin of the proton and the electron is always either $+1$, -1, or 0, depending on how you pick the signs. The proton and electron together can *never* have a total spin of either $+\frac{1}{2}$ or $-\frac{1}{2}$, as the original neutron had. This means that the law of conservation of angular momentum (another very familiar and rigidly observed law of conservation) is broken.

If, however, the neutron breaks down into a proton, an electron, and a neutrino, all three of which have spins of either $+\frac{1}{2}$ or $-\frac{1}{2}$, and the sum of the three ($+\frac{1}{2}$, $+\frac{1}{2}$, $-\frac{1}{2}$, for instance) adds up to $+\frac{1}{2}$, which would be that of the original neutron, angular momentum is conserved.

There is a fourth law of conservation, discovered much later than the others: the law of conservation of lepton number. A neutron and a proton each have a lepton number of 0, while an electron has a lepton number of $+1$ and a positron a lepton number of -1.

A neutron, then, starts off with a baryon number of $+1$ and a lepton number of 0. If it breaks down into a proton (baryon number equal to $+1$ and a lepton number equal to 0) and an electron (baryon number equal to 0 and a lepton number equal to $+1$), the baryon number is conserved, but lepton number is *not*.

But suppose a neutrino is also formed with a lepton number of -1. In this case, the neutron (baryon number $+1$ and lepton number 0) breaks down into a proton (baryon number $+1$ and lepton number 0), an electron (baryon number 0 and lepton number $+1$), and a neutrino (baryon number 0 and lepton number -1). Here, you start with a neutron (baryon number $+1$ and lepton number 0) and end with three particles with a total baryon number of $+1$ and a total lepton number of 0. Lepton number is now conserved, as well as baryon number.

Of course, for the neutrino to have a lepton number of -1, it should by rights be the mirror image (antineutrino) of the particle, but this is all right. An antineutrino conserves the laws of energy, electric charge, momentum, and angular momentum just as well as a neutrino. The antineutrino also conserves lepton number.

An undetectable particle designed simply to save a single law of conservation is not very convincing. Four different undetectable particles designed to save each of four laws of conservation are even less convincing. However, a single undetectable particle that happens, in itself, to save each of four conservation laws—energy, momentum, angular momentum, and lepton number—becomes very convincing. As the years went by, physicists, more and more, took the attitude that neutrino and antineutrino, whether detected or not, must exist.

Detecting the Antineutrino

Physicists would not feel completely comfortable about the existence of the neutrino and antineutrino until they had been detected. (Usually, the term neutrino is used, for simplicity's sake, to include the antineutrino.)

To be located, a neutrino must interact with another particle, and the interaction must be detectable, as well as distinct from other interactions. In other words, you must recognize an interaction as being caused by a neutrino and by nothing else.

This is not easy because the neutrino hardly ever interacts with anything. The average neutrino can pass through 3,500 light-years of solid lead, it is calculated, before being absorbed.

This is the *average* neutrino. Individual neutrinos might continue avoiding direct hits by sheer chance and travel twice as far as that, or a million times as far, before being absorbed. Others might just happen to make a direct hit and be absorbed after having traveled only half the average distance, or one-millionth. This means that if you deal with a beam containing trillions upon trillions of neutrinos and have them all pass through a quantity of matter in the laboratory, a very few might just happen to strike some particle in that matter and interact.

To have any chance of detecting a neutrino, then, you must have a rich source of them. That rich source became available once nuclear reactors, based on fissioning uranium nuclei, were devised.

Uranium nuclei, being very complex, need a great many neutrons even to be nearly stable; 143 neutrons to 92 protons, in the case of uranium-235. When the uranium nucleus breaks into two smaller fragments, each requires fewer neutrons to be stable, and so some neutrons are set free. With time, many of these break down into protons, liberating antineutrinos as well. A typical fission reactor might easily give off a billion billion antineutrinos every second.

The next problem would be to decide what to expect the antineutrino to do. We know that a neutron breaks down into a proton by emitting an electron and an antineutrino. Can we reverse this process by having a proton absorb an electron and an antineutrino at the same time to become a neutron again?

This is asking a great deal. An antineutrino hitting a proton is very unlikely in itself. To expect it to hit a proton just as an electron is also hitting the same proton is asking entirely too much. It would happen so rarely as to be quite an impractical process. However, having an electron hit a proton is the equivalent of having a proton emit a positron. (This is like saying that having someone give you a dollar is equivalent to having someone pay off a dollar debt that you might have. Either way, your assets go up by one dollar.)

This means that it is possible for an antineutrino to hit a proton, which will then give up a positron and become a neutron. This conserves the baryon number because a proton becomes a neutron, both with a baryon number of $+1$. It also conserves the lepton number, for an antineutrino disappears and a positron appears, each with a lepton number of -1. It conserves electric charge because a proton disappears and a positron appears, each with a change of $+1$. The laws of conservation of energy, momentum, and angular momentum are also upheld.

Suppose, then, that an antineutrino does hit a proton and produces a neutron and a positron. How can you tell that it has happened? It only happens at long intervals and, meanwhile, all sorts of other interactions are taking place, drowning out the antineutrino action. But if a neutron and a positron are produced, the positron is bound to combine with any electron it meets within a millionth of a second, undergoing mutual annihilation. In the process, two gamma rays are formed of equal strength, traveling in opposite directions, with a total energy equivalent to the mass of the two particles. As for the neutron, it could be quickly absorbed by the nucleus of a cadmium atom (if there are any in the neighborhood). The nucleus will gain enough energy in the process to emit three or four photons with a fixed total energy.

No other known interaction will produce exactly this result. If, then, you locate photons emitted all at the same time, in the proper directions, and with the proper energies, you have detected the interaction of a neutrino with a proton, and nothing else.

A team of American physicists led by Frederick Reines (b. 1918) and Clyde Lorrain Cowan (b. 1919) began to tackle the problem along these lines in 1953. They had the use of a fission reactor and saw to it that as many of the antineutrinos as possible struck large tanks of water, which contained trillions of trillions of protons in the hydrogen and oxygen nuclei making up the water molecules. The chemical cadmium chloride had been dissolved in the water. The cadmium nuclei would act to pick up any of the neutrons given off. Finally, the setup included devices that would detect the gamma ray photons and determine their directions and energies, so that all they had to do was to wait for the right combination of photons.

Naturally, in order to detect the right combination as easily as possible, they had to cut out as many wrong combinations as possible, so they shielded the entire apparatus more and more efficiently as time went on. In the end, virtually nothing got in *but* antineutrinos. Eventually, they managed to cut down enough of the "background noise" to be sure that they were detecting the occasional interaction of an antineutrino with a proton. In 1956, twenty-six years after Pauli's suggestion, Reines and Cowan announced that they had detected the antineutrino.

Other scientists instantly tried to repeat the experiment, or to try modifications, and there was no question about it: given the proper equipment, anyone could detect antineutrinos. It was no longer a ghost particle, but exactly the particle that Pauli and Fermi had deduced as being necessary to explain the details of beta-ray radiation and the breakdown of the neutron. (What a testimonial this was to the value of the use of logic in science. It also showed how important it is to stick to a good theory—such as the various laws of conservation—as long as is reasonably possible. Of course, the time might come when an idea that has seemed as firm as steel must be given up—even laws of conservation—and we shall come across such cases. It is the glory of science that it occasionally corrects itself, however reluctantly. No other variety of human intellectual endeavor seems to have quite the same built-in machinery with which to do so.)

Detecting the Neutrino

Just as fission produces a flood of antineutrinos because of the numerous conversions of neutrons to protons it makes necessary, so fusion produces a flood of neutrinos because of equally massive conversions of the same two particles. In the fusion of hydrogen to helium, for instance, four hydrogen nuclei, made up of four protons altogether, are converted into one helium nucleus made up of two protons and two neutrons. In the process, two positrons are formed, and with the positrons, two neutrinos.

Whereas we have working fission reactors to provide

235

us with floods of antineutrinos, we do not, as yet, have working fusion reactors to provide us with floods of neutrinos. The uncontrolled fusion of a hydrogen bomb produces floods of neutrinos for a while, but working close enough to such an explosion in order to take advantage of the burst is not a very practical notion. However, we do have an enormous and continuously "exploding" hydrogen bomb about 93 million miles away: the Sun. It produces an incredible number of neutrinos every second, and has been doing so for about four and a half billion years.

Produced at the center of the Sun, where fusion is taking place, are photons. The photons react very readily with matter, so that they are absorbed, reemitted, absorbed again, and so on indefinitely. It takes a tremendous length of time for photons to make their way from the Sun's center to its surface, where they are launched into space, some to reach the Earth. So much has happened to these photons in their travels within the Sun, however, that nothing about them is likely to tell us very much about what happens at the Sun's center.

Neutrinos are another case. They interact so slightly with matter that, after being formed, they move from the Sun's center to its surface in a little over two seconds. (Because neutrinos are massless, they travel at the speed of light, as do photons and gravitons.) Once the neutrinos reach the Sun's surface, they continue going and, if they happen to be moving in the direction of the Earth, reach us in eight minutes.

Because the neutrinos reach us directly from the Sun's center, there is at least a chance that, from their properties, we might be able to garner information about that center that is unavailable in other ways. To detect solar neutrinos, then, is much more than to simply prove their existence. It is to investigate the Sun.

In order to detect neutrinos, we have to make use of

a particle interaction that is the reverse of the one used to detect antineutrinos. To detect antineutrinos, we have them strike protons to produce neutrons and positrons. To detect neutrinos, we must have them strike neutrons to produce protons and electrons. In detecting antineutrinos, then, we need a target rich in protons, such as water. In detecting neutrinos, we need a target rich in neutrons, and for that we have to use a neutron-rich nucleus.

A particularly neutron-rich nucleus suggested by the Italian physicist Bruno M. Pontecorvo (b. 1913) is chlorine-37, which has a nucleus made up of 17 protons and 20 neutrons. If chlorine-37 absorbs a neutrino, one of its neutrons will be turned into a proton, with the emission of an electron.

But why this neutron-rich nucleus rather than any other? Because when the chlorine-37 nucleus loses a neutron and gains a proton, it becomes argon-37 (18 protons and 19 neutrons), which is a gas that can be easily removed from the material containing the chlorine-37 nuclei. The recovery of such a gas indicates a neutrino absorption and nothing else.

It might seem that the best way of getting a chlorine-37 target is to use chlorine itself, but chlorine is a gas, and it would be difficult to separate very small quantities of another gas from it. One might liquefy the chlorine (and argon would still be a gas at the temperature of liquid chlorine), but that would require refrigeration. It would be much better to use a compound that is liquid at room temperature and that has molecules containing many chlorine atoms.

One such compound is perchloroethylene, in which each molecule is made up of two carbon atoms and four chlorine atoms. It is a chemical used as a common dry-cleaning compound, and is not very expensive. If even a few atoms of argon-37 are formed, they can be flushed out of the liquid by a stream of helium, and can then be detected, for argon-37 is radioactive and can be identified even in minute traces by its characteristic form of breakdown.

It was by using this interaction that the American physicist Raymond E. Davis showed that the neutrino actually exists.

Beginning in 1965, Reines, one of the discoverers of the antineutrino, began working on the detection of solar neutrinos. He made use of large vats of perchloroethylene buried deep in a mine, with a mile or so of rock between it and the surface. The rock would absorb all radiation, even cosmic-ray particles, *except* for neutrinos, which could easily pass through the entire Earth. (There might also be some particles originating from radioactive material in the rocks immediately surrounding the experiment.)

It is odd to think that the Sun must be studied from a vantage point a mile underground, but that is what Reines proceeded to do. However, whatever way he tried to improve his techniques and to refine his instruments, Reines never obtained more than about ⅓ of the neutrinos he expected to detect.

Why? It might have been that the nature of the observations were in some ways inadequate; or that we don't know all there is to know about neutrinos; or that our theories about what is going on at the center of the Sun are mistaken. No solution has yet been reached on this "mystery of the missing neutrinos," but, when it comes, it is sure to be exciting.

If the Sun produces neutrinos, so, we may be sure, do other stars. However, even the nearest star, Alpha Centauri (a system containing two Sun-like stars, and one dim dwarf star), is some 270,000 times as far from us as our Sun. The number of neutrinos reaching us from the Alpha Centauri stars can be, at best, about one fifty-billionth as many as those supplied us by the Sun. We can just barely detect the neutrino emission of the Sun, so we haven't the slightest chance of detecting any from other, normal stars.

But not all stars are normal. Every once in a while, a star explodes as a gigantic supernova, in which radiation of all types suddenly increases a hundred-billion-fold.

In February 1947, such a supernova appeared in the Large Magellanic Cloud, 150,000 light-years away. It was 33,000 times as far away as Alpha Centauri, but the floods of neutrinos it produced more than made up for that. It was the closest supernova to us in nearly 400 years, and the first for which we had "neutrino telescopes," such as that Reines had been working with.

One such neutrino telescope was placed under the Alps. A team of Italian and Soviet astronomers there detected a sudden burst of seven neutrinos the night before the supernova was detected by eye. It turns out, then, that as astronomers improve their ability to detect and study neutrinos, they will not only learn more about what goes on in the center of our Sun, but also about what goes on in colossal star explosions and, perhaps, about other facets of astronomical lore.

Other Leptons

We have thus far described four leptons: the electron, positron, neutrino, and antineutrino. The electric charges are, respectively, -1, $+1$, 0, and 0. The masses are (if we set the mass of the electron at 1), respectively, 1, 1, 0, and 0. Their spins are either $+\frac{1}{2}$ or $-\frac{1}{2}$; it is this half-spin that makes them all fermions. (Photons and gravitons have 0 mass and 0 electric charge, but they have spins of 1 and 2, respectively; the whole-number spin makes them bosons.)

This was the situation as late as 1936, when the neutrino, the antineutrino, and the graviton had not yet been actually detected, but seemed certain of existence in each case. At the time, there were also four baryons known: the proton, neutron, antiproton, and antineutron.

Add to these the photon and graviton and you have ten particles altogether, which seemed to account for every piece of matter in the Universe, as well as all of the interactions that scientists had observed. It would have been nice to end there, for a ten-particle Universe is reasonably simple.

In 1936, however, Anderson, who had discovered the positron four years earlier, was still studying cosmic rays in the mountains, and noticed particle tracks that curved in an odd manner. The curve was less sharp than that of an electron, so it had to be more massive than an electron (assuming that the new particle had the same electric charge). It was sharper than that of a proton, however, indicating that it had to be less massive than a proton. Moreover, there were curves of this sort, otherwise exactly alike, in both directions, indicating that some were particles and some antiparticles.

The conclusion was that there existed particles and antiparticles of intermediate mass, between those of the known leptons and the known baryons. Measurements showed that the new particles were 207 times as massive as electrons and, therefore, about ⅑ the mass of a proton or neutron.

Anderson at first called the new particle a mesotron, the prefix meso- coming from a Greek word meaning "middle," or "intermediate." Notice again that improper -tron ending. This time, the ending did not stick, fortunately, and the term *meson* came to be used, instead, as a general term for all particles of intermediate mass. Because Anderson's particle turned out, eventually, to be only one of a number of such intermediate-mass particles, each had to be distinguished from the others. Anderson's particle came to be called a mu meson, where mu is a letter of the Greek alphabet, equivalent in sound to the English *m*. Then it turned out, as I shall explain later, that the mu meson was different from other intermediate-mass particles in a very fundamental way. The term meson was therefore restricted to the others and *did not include the mu meson*. Anderson's particle is, therefore, now called a muon.

The muon was the first particle discovered that did not have an obvious use either in forming part of the structure of atoms, in preserving the laws of conservation, or in mediating subatomic interactions. The Austrian-American physicist Isidor Isaac Rabi (1898–1988) is supposed to have asked, on hearing of the muon, "Who ordered that?"

The muon has an electric charge of -1, precisely that of the electron, whereas the antimuon has an electric charge of $+1$, precisely that of the positron. In fact, the negative muon, except for its mass and one other property, is in all respects identical to the electron, while the antimuon is similarly identical to the positron. This is true of such things as electric charge, spin, and magnetic field. A negative muon can even replace an electron in an atom, producing a short-lived muonic atom.

In order to conserve angular momentum, a muon must have the same angular momentum as the electron it replaces. In that the muon has much more mass than an electron, which would add to the muon's angular momentum, it must decrease it by moving in closer to the nucleus than the electron ever is. We can also see that this must be so because the muon, with far greater mass than the electron, has a much shorter associated wave, which can squeeze into a much tighter orbit.

Because muonic atoms are, for these reasons, much smaller than electronic atoms, two muonic atoms can get much closer to each other than electrons can. The nuclei of

241

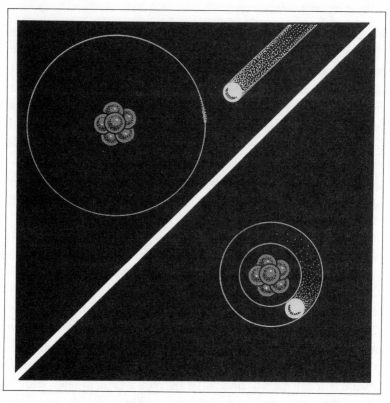

A negative muon, except for its mass and one other property, is identical to an electron. It can even replace an electron in an atom, producing a short-lived muonic atom.

muonic atoms, therefore, have a far greater tendency to fuse than do ordinary electronic atoms. Muonic atoms might thus seem a possible route for practical fusion—except for one enormous catch that we will come to later.

Muons and antimuons can undergo mutual annihilation, producing 207 times as much energy as do electrons and positrons. Similarly, if 207 times as much energy is concentrated into a tiny area as would suffice to form an electron-positron pair, a muon-antimuon pair can be formed.

But what happens if a muon is produced without its

antimuon, as an electron is produced without a positron by the breakdown of a neutron? There are negatively charged particles more massive than a muon (particles I will get to later), which break down to form a muon *without* also forming an antimuon. Similarly, there are positively charged particles that will form an antimuon without a muon.

This does not violate the law of conservation of electric charge, but, as in the case of neutron breakdown, there is a violation of the laws of conservation of energy and momentum. In addition, the heavier particles that break down into muons are themselves neither baryons nor leptons, but the muon *is* a lepton; therefore, while the breakdown does not violate the law of conservation of baryon number, it does violate the law of conservation of lepton number by seeming to form a lepton out of nothing. Here, too, as in the case of neutron breakdown, the simultaneous formation of neutrinos and antineutrinos preserves the laws of conservation, but with an added complication I will get to later.

We can think of a muon as merely a massive electron, and of the antimuon as a massive positron, but why should they exist, and why should they have masses equal to 207 times that of an electron or positron, rather than any other mass?

Let's make an analogy. Suppose an electron is a golf ball lying at the bottom of an energy valley. There is no way for it to descend any lower, so it just stays there. If, however, energy is added to it (as would happen if a golf ball in that position were struck by a golf club), the added energy would cause the golf ball to roll up the hillside. It would reach some maximum height, then roll down to the valley bottom again, giving up the added energy as it did so.

The harder the golf ball is hit, the higher up the hillside it rolls before descending again. If it is hit hard enough, it might just roll high enough to reach a ledge on the hillside

243

where it might remain. It would still be a golf ball (i.e., electron), but it would have gained sufficient energy to be far above its normal position at the bottom of the valley, or, in subatomic terms, it would have gained far more mass than it had at the bottom of the valley.

The ledge happens to be at an energy height equivalent to 207 times the mass of the electron. Why is it just at that height? We don't know, but our inability to advance an exact reason is not dreadfully disturbing. (Science has not succeeded in explaining everything about any significant branch of knowledge and, perhaps, never will. Scientists have discovered innumerable answers, or apparent answers, to innumerable problems, but each answer supplies us with new and more subtle, and, perhaps, more intractable problems.)

Unstable Particles

The electron, as we know, is a stable particle. By stable, I don't mean that nothing can happen to it. If an electron meets a positron, both undergo annihilation and are converted into photons. If an electron collides with particles other than a positron, it can undergo other types of changes.

However, if an electron is isolated in space and does not encounter any other particles, it will (as far as we know) remain in existence, retaining all of its properties unchanged, forever.

The same is true of a positron, as well as of a neutrino and an antineutrino. All four leptons known prior to the discovery of the muon are stable particles. (The same is true of the two bosons known in the 1930s, the photon and the graviton.)

Of the particles that are *not* leptons, and that were

known prior to the discovery of the muon—the proton, antiproton, neutron, and antineutron—the proton and antiproton seem to be stable (although there are now doubts about this as we shall see).

The neutron and the antineutron are *not* stable. If a neutron is isolated from all other particles, it will nevertheless break down into a proton, an electron, and an antineutrino, while an antineutron will break down into an antiproton, a positron, and a neutrino. However, this is a relatively slow process, taking several minutes on average. In addition, when a neutron forms part of a nonradioactive nucleus, it is stable and can remain there, unchanged, indefinitely. Muons, however, break down into electrons almost at once. The average muon, when left strictly to itself, breaks down in only $\frac{1}{500,000}$ of a second.

Why should the muon endure for so short a time? Consider the analogy I have already used, in which an electron is driven up the hillside of mass and comes to rest on a ledge representing 207 times the mass at the bottom of the valley. We can picture the ledge as a narrow one, and the muon resting on that ledge as vibrating, or trembling. As a result of this tremble, sooner or later, the muon falls off the ledge, slides down to the valley below, becoming an electron again. The "sooner or later" turns out, from the narrowness of the ledge and the magnitude of the tremble, to be $\frac{1}{500,000}$ of a second.

All objects, including you and me, exhibit a type of tremble, dictated by the fact that quantum mechanics shows all objects to be associated with a wave aspect. For ordinary objects, the tremble is so excessively minute as to be of no importance, but the smaller the mass, the more marked the tremble relative to the size of the object. Subatomic particles have so little mass that the tremble gains considerable importance and must be taken into account in any study of their properties.

The electron also has a tremble—one even more marked than that of the muon—but the electron is at the bottom of the valley. It has no way of falling any farther, and is therefore stable.

In 1975, the American physicist Martin Perl detected an electron-like particle, even more massive than the muon, in the debris produced by collisions within accelerators. He called this particle a tau lepton, where tau is a letter in the Greek alphabet equivalent to the English *t*. It is also called a tauon.

The tauon shares all of its properties but two with the electron and the muon. Of the two properties that are distinctive, one is mass. The tauon is a super-massive electron, with a mass about 3,500 times that of an electron, and nearly 17 times that of a muon. It is nearly twice as massive as a proton or a neutron, and yet from the way it behaves it is clear that it is a lepton, even though that name is used chiefly for the far better known particles of little or no mass. (It might sound confusing and contrary to common sense that a name that implies smallness should be applied to a very massive particle, but consider this as an analogy. Reptiles, including alligators, anacondas, and the extinct dinosaurs, are *much* larger than insects, if one considers each group as a whole. There are, however, Goliath beetles as large as your fist, while there are lizards small enough to fit on your fingertip; yet the Goliath beetle is an insect and the lizard is a reptile.)

The second way in which a tauon is distinct is in its instability. It is far more unstable than the muon, for it breaks down in only a five-trillionth of a second. In doing so, it changes into a muon, which, in turn, changes into an electron.

We can imagine the tauon resulting from a gain in energy that drives it far higher up the hillside of mass. It reaches a much higher and much narrower ledge than the muon attained. The tauon remains on that ledge for the barest instant before falling off. Can we expect to find additional leptons, then, more massive and more unstable than the tauon? Are there an infinite number of ledges on our allegorical hillside, each higher and narrower than the one before?

Apparently not. Physicists have reason to believe, on the basis of some recent, rather involved observations, that three is the limit and that we now have located all of the leptons there are.

Tauonic atoms, if they existed, would be even smaller than muonic atoms, and therefore still easier to fuse. By now, though, you probably see the catch. These heavy leptons are far too unstable to serve as practical routes to fusion. They would be gone before we could do much with them.

Neutrino Varieties

Let's go into the breakdown of the muon in some detail, because there's a problem there involving the laws of conservation. Suppose the muon breaks down into an electron and an antineutrino. This conserves electric charge and momentum. However, angular momentum is *not* conserved. The muon has a spin of $+\frac{1}{2}$ or $-\frac{1}{2}$, which is also true of the electron and the antineutrino. An electron and an antineutrino, taken together, have a spin of either $+1$, 0, or -1, depending on the signs of their spins. The two cannot possibly have a total spin of either $+\frac{1}{2}$ or $-\frac{1}{2}$.

Why does the muon breakdown seem to break the law of conservation of angular momentum, even with an antineutrino taken into account, when the neutron breakdown does not? This is because the neutron breaks down into *three* particles—a proton, electron, and antineutrino—and three half-values can add up to a total half-value. The muon breakdown as we have described it so far produces only two particles—an electron and an antineutrino—and two half-values can add up to only an integral value, never a half-value. In other words, we have to suppose that the muon, upon breaking down, also produces three particles— an electron and *two* antineutrinos perhaps.

Unfortunately, this does not necessarily mean that conservation is preserved. The muon has a lepton number of +1. The electron and the antineutrino each have a lepton number of +1, so that we begin with +1 and end with +2, violating lepton number conservation. If we add a second antineutrino, we start with +1 and end with +3, which is still worse. However, suppose the muon breaks down into an electron, an antineutrino, and a neutrino. The neutrino has a lepton number of −1, so the three particles that are produced have lepton numbers of +1, +1, and −1, for a total of +1, which is the lepton number of the original muon. If, then, we suppose that a muon, in breaking down into an electron also produces both an antineutrino and a neutrino, all of the laws of conservation are preserved.

Happy ending? Yes, except for one small point. All of the neutrino-producing interactions physicists had, until the discovery of the muon, observed produced *either* a neutrino *or* an antineutrino. Muon breakdown is odd in that it produces *both* a neutrino *and* an antineutrino.

Can it be that there are two types of neutrino? Can it be that one is produced only by electrons and the other only by muons, so that one can speak of an electron neutrino (and antineutrino) and a muon neutrino (and antineutrino)? Can it be that when the muon breaks down into an electron the muon and electron each produce a variety of neutrino and that this is why the muon breakdown involves two neutrinos? This is called the two-neutrino problem.

If an electron neutrino and a muon neutrino are different in nature, this must be because of some difference in properties—but physicists have not been able to find any. It is even more difficult to study the neutrinos produced by muons than those produced by electrons, but as nearly as physicists can tell, the two types of neutrino are identical. Both have 0 charge, 0 mass, either $+\frac{1}{2}$ or $-\frac{1}{2}$ spins, and so on.

Does this settle the question? Of course not. It might well be that there is a difference in some respect that no scientist has ever thought of, and therefore that no scientist has ever tried to measure, assuming we have the devices with which to measure it.

But if we can't spot any difference directly, perhaps we can spot one indirectly by having the particles make the judgment themselves. Suppose, for instance, that an antineutrino produced by an electron meets a neutrino produced by a muon. If they are identical in all respects, except for being mirror images, they ought to undergo mutual annihilation and produce a minute pulse of energy. If they differ in *any* respect other than being mirror images, they should *not* undergo mutual annihilation. If there is no annihilation, the particles recognize a difference between themselves, and that is good enough. We will take their word for it, even if we don't know what the difference consists of.

However, neutrinos are such infinitely minute, nonreactive particles, that the chance of two of them encountering each other is essentially zero. Even if they did, the energy produced might well be too small to be observed. There is, however, another way in which neutrinos can be made to reveal their nature. If an electron produces only electron neutrinos and a muon produces only muon neutrinos, if the interactions are reversed, the electron neutrino should bring about the production only of electrons and the muon neutrinos only of muons. If the two neutrinos are truly identical, however, they should produce electrons and muons in equal numbers.

Such an experiment was carried out in 1961 by a team headed by the American physicist Leon Max Lederman (b. 1922). They began by hurling high-energy protons into a target of the metal beryllium. This produced a vast number of particles, among which were high-energy muons that decayed to produce high-energy muon-neutrinos. This vast melange of particles was then hurled at a 12-meter-thick slab of steel, which absorbed everything but the neutrinos (which can pass through anything). On the other side of the steel, the stream of energetic muon neutrinos entered a device that could detect neutrino collisions. There wouldn't be many, of course, but over a period of eight months, 56 such collisions were noted, and every one of them produced a muon.

This made it clear that a muon neutrino could *not* produce an electron and was therefore different in some distinctive way (whether we knew the way or not) from an electron neutrino. Lederman got a Nobel prize for this in 1988.

Lederman's work meant that the conservation of lepton number is a little more complicated than had been thought. There is a conservation of electron number and a conservation of muon number separately. Thus, an electron has an electron number of +1, and a positron −1. An electron neutrino has an electron number of +1, and an electron-antineutrino −1. All four have a muon number of 0. In the same way, a muon has a muon number of +1, and an antimuon −1. A muon neutrino has a muon number of +1, and a muon antineutrino −1. All four have an electron number of 0.

When a muon, with a muon number of +1 and an electron number of 0, breaks down, it forms an electron

(electron number +1, muon number 0), an electron anti-neutrino (electron number −1, muon number 0) and a muon neutrino (electron number 0, muon number +1). All three breakdown particles added together have a muon number of +1 and an electron number of 0, which is what is true of the original muon. The breakdown, therefore, conserves both electron number and muon number.

The tauon also produces a neutrino, which has been little studied so far, that physicists suspect has all of the obvious properties of the other two neutrinos but that is somehow distinct from them. It seems unavoidable to suppose that there is such a thing as conservation of tauon number.

Physicists now speak of three "flavors" of leptons. These are (1) the electron and the electron neutrino, (2) the muon and the muon neutrino, and (3) the tauon and the tauon neutrino. There are also three flavors of antileptons: (1) the antielectron (positron) and the electron antineutrino, (2) the antimuon and the muon antineutrino, and (3) the tauon and the tauon antineutrino. (The term *flavor* is, in some ways, an unfortunate one. It is used in common English to distinguish objects by taste, as in different flavors of ice cream. It isn't quite right to give nonscientists the notion that the differences between subatomic particles are "shades" of differences rather than absolute, measurable ones. However, scientists are human and sometimes reach for a dramatic or even humorous term. For example, some atomic nuclei are easier to hit with subatomic projectiles than others. Those nuclei that are particularly easy to hit were said by some whimsical scientists to be as easy to hit "as the side of a barn." As a result, the nuclear cross-section that gives the measurement of the ease with which a particular nucleus can be targeted is given in a unit called a barn.)

There are, then, twelve leptons and antileptons all

together. They are fundamental particles (or at least are currently considered such) because they will not spontaneously break down into particles that are simpler than leptons. The tauon and the muon break down into electrons, whereas the antitauon and the antimuon break down into positrons. The electron, the positron, the three neutrinos, and the three antineutrinos don't seem to break down at all.

Why are there twelve leptons, when the Universe seems to contain only electrons and electron neutrinos in appreciable numbers? Electron antineutrinos are produced only in radioactive transformations, which are few in number in the Universe as a whole. Positrons are produced in some radioactive transformations, but less often even than electron antineutrinos. The heavier leptons and their neutrinos are produced, as far as we know, only in the laboratory by such things as cosmic-ray bombardment.

Why, then, doesn't the Universe get by on just electrons and electron neutrinos? Why needlessly complicate things? Because, my instinct tells me, the complications are not needless. The Universe is built in such a way that every interaction must play its role. We might not see what possible use the tauon has, for instance, but I have the strong feeling that whatever it is that makes our Universe work as it does requires the tauon's existence; that without the tauon the Universe would not be the Universe we live in and might not even have the capacity to exist.

10

INTERACTIONS

The Strong Interaction

Putting leptons to one side, what about the baryons? What about the particles that make up the atomic nucleus? These represented a serious problem once the neutron was discovered and the proton-neutron structure of the nucleus was advanced. The problem is summed up in the following question: "What holds the nucleus together?"

Until 1935, there were only two interactions known that could hold objects together: the gravitational and the electromagnetic. Of these, the gravitational interaction was so weak that it could be entirely disregarded in the world of subatomic physics. It only makes itself seriously felt when a huge mass is accumulated. It is important on the level of satellites, planets, stars, and galaxies, but certainly not where atoms and subatomic particles are concerned.

That left the electromagnetic interaction. The electromagnetic attraction between positive and negative electric charges is perfectly adequate to explain how molecules are held together in crystals, how atoms are held together in molecules, and how electrons and nuclei are held to each other in atoms, but when scientists got all the way down to the atomic nucleus, they encountered a problem.

As long as they thought that nuclei were made up of protons and electrons, there seemed no problem. The protons and electrons attracted each other strongly; the more strongly, in fact, the closer together they were. In the nucleus, they were virtually in contact. The protons were also virtually in contact with each other, as were the electrons. Between objects of the same electric charge there was a repulsion just as strong as the attraction between objects of opposite charge.

Within the nucleus, then, it might be expected that the protons would repel each other and the electrons would also repel each other, but, presumably, they were intermingled and placed in such a way that the attractions were more effective than the repulsions. This was true in crystals, which were often built up of an intermixture of positively charged ions and negatively charged ions, with the opposite charges distributed so that attractions overwhelmed repulsions and the crystals hung together. In short, the electrons within the nucleus acted as a type of cement for the protons, and vice versa. Between the action of the two cements, the nucleus held together.

But the nature of the nuclear spin and the necessity of conserving angular momentum cast serious doubt on the validity of the proton-electron theory of nuclear structure. With the discovery of the neutron, it quickly became obvious that it was necessary to assume a proton-neutron structure that would solve all of the difficulties raised by the proton-electron structure—except one. The cement had disappeared.

If we consider the electromagnetic interaction only, then, the only force that could make itself felt inside a proton-neutron nucleus is the extremely strong repulsion of each proton for all others. The neutrons, being electrically uncharged, neither attract nor repel the protons and are merely innocent bystanders, so to speak. The strong proton-proton repulsion should suffice to explode the nucleus instantly into individual protons.

Yet this does not happen. The atomic nucleus remains quietly in place, quite stable, with no sign of mutual destructive repulsion among protons. Even in the case of those nuclei that are radioactive, explosions take place in a strictly limited fashion, turning a proton into a neutron or vice versa, eliminating a two-proton, two-neutron alpha particle, or, in extreme cases, breaking into two halves. All of this happens relatively slowly, sometimes exceedingly slowly. *Never* does any nucleus explode, instantaneously, into individual protons.

The natural conclusion we must come to is that there is some interaction involved that is neither gravitational nor electromagnetic—one that human beings have never thought of, let alone studied—and that it is this interaction that holds the nucleus together. It might be called a nuclear interaction.

The nuclear interaction, whatever it is, must introduce a strong attractive force—one that is far stronger than the repulsive force produced by the positive charges on the various protons. In fact, as it eventually turned out, the nuclear interaction produces an attractive force over 100 times as intense as that of the electromagnetic interaction. This is, in fact, the strongest force that is known to exist between subatomic particles (and, it is thought, the strongest that can exist). It is therefore usually called the strong interaction. But what is the strong interaction? How does it work?

The first to consider the strong interaction was Heisenberg, who, in 1932, had first suggested the proton-neutron structure of the nucleus. This is not surprising. When a scientist advances a startling notion that solves a great many problems, but which has a gaping hole in it, he is bound to do his best to mend that hole. After all, it's his baby.

Heisenberg worked up the notion of exchange forces. This is something that classical physics, the type of physics that existed before quantum mechanics was devised, could not deal with or understand. If quantum mechanics is used, however, exchange forces are seen to be possible, and effective.

To explain it without mathematics, we can imagine protons and neutrons constantly exchanging something. Let us say that what they exchange (as Heisenberg first suggested) is electric charge. This means that the positive electric charges inside a nucleus are constantly being transferred from a particle possessing a positive charge to a particle that does not. This means that every baryon would be a proton and a neutron in exceedingly rapid alternation. No proton can feel repulsion because before it has time to react to the repulsion it is a neutron. (It's like bouncing a hot potato rapidly from hand to hand to keep from being burned.)

Such an exchange force would set up a powerful attraction and keep the nucleus together; but on closer examination Heisenberg's suggestion proved inadequate, unfortunately. Then the Japanese physicist Hideki Yukawa (1907–1981) tackled the job. It seemed to him that if exchange forces worked inside the nucleus for the strong interaction, they would have to work for all interactions. He

applied quantum mechanics to the electromagnetic interaction and, it appeared to him, that what was exchanged
was a particular particle, the photon. It was the continual,
extremely rapid exchange of photons between any two particles with electric charge that produced the electromagnetic interaction. Between particles with the same charge,
the exchange produced a repulsion, and between those with
opposite charges, it produced an attraction.

Between any two particles with mass, there was also
a rapid exchange of gravitons. (These particles have never
been detected because they are so weakly energetic that
nothing we have yet developed is sensitive enough to demonstrate their existence unequivocally; however, no physicist doubts their existence.) Since there seems to be only
one type of mass, the gravitational interaction produces
only an attraction.

Within the nucleus, then, there would have to be another such exchange particle that dashed endlessly between
the protons and neutrons within the nucleus. Here, however, there was a difference. Both the electromagnetic and
gravitational interactions were long-range effects whose intensity declined only slowly with distance. The effect of
electromagnetism, possessing both attraction and repulsion, is muted, but we can see clearly what this means in
connection with gravitation where masses can be huge and
where there is *only* attraction. The Earth holds the Moon
firmly, even at a distance of nearly 400,000 kilometers
(237,000 miles). The Sun holds the Earth firmly, at a distance of 150,000,000 kilometers (93,000,000 miles). Stars
hold together in galaxies and galaxies in clusters over thousands and even millions of light-years.

The strong interaction, however, decreases in intensity
with distance *much* more rapidly than does either gravitation or magnetism. Double the distance, and the latter
two interactions decrease to one-fourth their intensity;

however, double the distance, and the strong interaction decreases to less than 1 percent of its intensity. This means that the strong interaction is very short-range indeed, and cannot be felt measurably except in the immediate vicinity of the particle that gives rise to it.

In fact, the effective range of the strong interaction is only about a ten-trillionth of a centimeter, or about $\frac{1}{100,000}$ the width of an atom. The only way, then, that protons and neutrons can feel the attractive effect of the strong interaction is to remain in close contact. This is why atomic nuclei are so small. They are just large enough to come under the umbrella of the strong force. Indeed, the largest nuclei known are so wide that the strong interaction has difficulty stretching the required distance, therefore there is a tendency for these nuclei to undergo occasional fission.

It is because of this difference in range that only gravitational and electromagnetic interactions are part of the common experience of human beings. The former has been known to exist from the very dawn of human intelligence, and the latter from the days of ancient Greece. The strong interaction, however, expressing itself only over nuclear distances, couldn't possibly be experienced until the nucleus was discovered and its structure understood—that is, not until the 1930s.

But *why* this difference in range? It seemed to Yukawa that, from quantum mechanical considerations, a long-range interaction required a massless exchange particle. In that the photon and the graviton, both massless, had electromagnetic and gravitational forces that were long-range, the strong interaction had to have an exchange particle with mass because it acted over a very short range. In fact, Yukawa calculated the mass of the exchange particle to be about 200 times that of the electron.

At that time, no particles were known with masses in this range, which gave Yukawa the depressing feeling that

258

his theory was wrong, but he published it anyway in 1935. Almost at once, however, Anderson discovered the muon and it was precisely in the range of mass that Yukawa had predicted for his exchange particle of the strong interaction. Naturally, everyone thought that the exchange particle had been discovered. Interest in Yukawa's theory flared up.

However, the interest died down quickly. The muon showed no tendency to interact with protons and neutrons, and thus could *not* be the exchange particle. In fact, it was simply not subject to the strong interaction at all, and that was the chief reason for classifying it as a lepton. None of the leptons are subject to the strong interaction. In fact, once the muon was recognized as merely a massive electron it was realized that it could no more be the exchange particle than was the electron.

Disappointment did not last forever. The British physicist Cecil Frank Powell (1903–1969) was studying the effect of cosmic rays on the atmospheric atoms and molecules they struck. He, too, like Anderson, invaded the mountain heights in order to do so. In the Bolivian Andes, for instance, he was at a high enough altitude so that the cosmic ray intensity (coming from outerspace and not absorbed by the lowermost thicknesses of Earth's atmosphere) was ten times higher than at sea level. Powell made use of special instruments of his own devising, more sensitive than Anderson's, to detect, in 1947, the curved tracks of particles of intermediate mass.

The new particle was judged, from its curvature, to have a mass equal to about 273 times that of an electron (close to Yukawa's prediction), being about a third more massive than the muon. The new particle was also just about as unstable as the muon, breaking down, on average, in about $\frac{1}{400,000}$ of a second.

These similarities were purely coincidental, however, for there was a deep and fundamental difference between the particles. Powell's particle was *not* a lepton. It was subject to the strong interaction and it interacted with protons and neutrons readily. In fact, it was the exchange particle that Yukawa had predicted.

The new particle was called the pi meson (pi being a Greek letter equivalent to the English letter *p*, standing for Powell, I presume). It was the first of a new class of particles, all subject to the strong force, called mesons (a name originally given to the muon, but withdrawn because the muon turned out to be a lepton). Despite the fact that the pi meson has the right to be called a meson, its name is frequently shortened for the sake of convenience to pion.

There is, of course, a positively charged pion, with a charge exactly equal to that of a proton or a positron, and a negatively charged antipion, with a charge exactly equal to that of an antiproton or an electron. The pion breaks down into a muon and a muon antineutrino, and the antipion breaks down into an antimuon and a muon neutrino, which conserves muon number because the muon and the muon antineutrino have muon numbers of $+1$ and -1, respectively, whereas the antimuon and the muon neutrino have numbers of -1 and $+1$, respectively. In that the pions have muon numbers of 0, the muon number is 0 both before and after the breakdown.

There is also a neutral pion, which has no electric charge and only about $29/30$ of the mass of the charged pions. It is less stable than the charged pions, has an average lifetime of only about a millionth of a billionth of a second, and breaks down into two gamma rays. The neutral pion is one of the few particles that, like the photon and the graviton, have no antiparticles; or, to look at it another way, it serves as its own antiparticle.

Mesons, by the way, have spins of 0, and are therefore not fermions; there is no law of conservation of meson number. Mesons can appear and disappear at will.

As a result of the discovery of the strong interaction, Yukawa received a Nobel prize in 1949, and Powell received one in 1950.

The Weak Interaction

While the strong interaction was an extremely dramatic discovery, it was not the first new interaction to be discovered in the 1930s. In 1933, Fermi, who was later to bombard uranium with neutrons (with consequences of enormous importance), grew interested in the work that Dirac had done on the electromagnetic interaction. In trying to describe the manner in which photons were emitted in electromagnetic interactions, Dirac had come upon the notion of antimatter.

It occurred to Fermi that the manner in which neutrons gave off electrons and neutrinos might be treated in the same way, mathematically, as was the manner in which particles gave off photons. He came to the conclusion that the mathematics worked, but that it indicated an interaction that was far different from the electromagnetic interaction that governed the release of photons. This new interaction, which was at first called the Fermi interaction, was much weaker than the electromagnetic interaction. It was, in fact, only about a hundred-billionth as intense as the electromagnetic interaction. (It was less than ten-trillionth as intense as the later-discovered strong interaction.)

The Fermi interaction was very short range, being felt only at distances equal to about a thousandth of the width of the atomic nucleus. It therefore played no role to speak of in the nucleus, but was of importance in the case of single particles. It was a second nuclear interaction (in the sense that it was a second short-range one involving only sub-

261

atomic particles). After Yukawa's theory was accepted, people spoke of the strong nuclear interaction and the weak nuclear interaction, the latter replacing the earlier Fermi interaction.

But then, in the interest of word economy, I suppose, the "nuclear" was dropped and scientists began to speak of the strong interaction and the weak interaction.

(This last is, to my way of thinking, not entirely appropriate, for though the weak interaction is far weaker than the strong and the electromagnetic reactions, it is, nevertheless, ten thousand trillion trillion times as intense as the gravitational interaction. It is the gravitational interaction that has a right to the name of weak.

(And yet I might be wrong in this. In that we are only truly acquainted with the gravitational interaction in connection with the huge mass of the Earth and of other astronomical bodies, there is no way in which we can consider gravitation weak in a practical, rather than a subatomic, sense. We have only to fall down to put weakness out of our mind in connection with gravitation. And, indeed, if enough mass is accumulated and squeezed into a small enough volume, the total gravitational intensity becomes stronger than anything conceivable—so intense that even the strong interaction could not fight it. If we stop to think of it that way, the weak interaction is the weakest, at least in the manner in which we ordinarily encounter it—so perhaps the name is not a bad one, after all.)

Some individual particles undergo changes, such as breakdowns, or interactions with each other, that are mediated by the strong interaction, and some undergo changes that are mediated by the weak interaction. Naturally, those events that are mediated by the strong interaction take place much more rapidly than those mediated by the weak interaction, just as a baseball travels more rapidly if thrown by a major-league pitcher than if thrown by a five-year-old child.

In general, events mediated by the weak interaction are likely to take place in a millionth of a second or so, while those by the strong interaction take less than a trillionth of a second—sometimes taking place in just a few trillionths of a trillionth of a second.

Baryons and mesons can both respond to either the strong interaction or the weak interaction, but leptons respond to the weak interaction only. (Baryons, mesons, and leptons can all respond to the electromagnetic interaction if, and only if, they are electrically charged. Neutrons, neutral pions, and neutrinos do *not* respond to the electromagnetic interaction.) This is why lepton events, such as the decay of the pion into a muon, that of the muon into an electron, or that of the radioactive productions of beta particles, all tend to happen in what seems slow motion on the subatomic scale. The neutral pion, which does not break down into muons, is affected by the strong interaction and therefore breaks down much more rapidly than do the charged pions.

The weak interaction differs from the other three in that it alone is not involved in some very obvious force of attraction. The gravitational interaction holds astronomical bodies together and allows the solar system to exist. The electromagnetic interaction holds atoms and molecules together and allows the Earth to exist. The strong interaction holds baryons together and allows atoms to exist.

The weak interaction does not hold anything together. It merely mediates the conversion of certain particles into other particles. This is not to be taken lightly, however. For one thing, it mediates the processes whereby protons join each other to form helium nuclei. This is the nuclear fusion process that keeps the Sun shining and makes life on Earth possible.

The weak interaction raised a problem, however. If the other three interactions all exert their effects by means of exchange particles, then the weak interaction must also have an exchange particle. In that the weak interaction is short-range, its exchange particle should have mass. In fact, in that the weak interaction is considerably shorter in range than the strong interaction, it ought to have an exchange particle considerably more massive than the pion.

A theory advanced first in 1967 (which I'll get to later) indicated that the weak interaction ought to have three exchange particles—one positively charged, one negatively charged, and one neutral—and that these must be perhaps 700 times as massive as the pion and 100 times as massive as the proton.

The exchange particles were referred to as W particles, the W standing for weak. The electrically charged particles were symbolized as W^+ and W^-, and the neutral particle Z^0.

It was important to find these exchange particles, not just to add them to the collection of particles scientists knew about, but because their existence would verify the theory that predicted them. This would be undeniably true if their masses were really as unbelievably enormous as the theory predicted. The theory was, as we shall see, an important one, and the detection of the exchange particles was vital to its importance.

The catch lay in the huge masses of the particles. Correspondingly huge energies would have to be disposed of to create particles that could be detected. It was not until 1955 that a large enough energy production capability became available to produce and detect the antiproton. To do the same for W particles would require the concentration of at least a hundred times as much energy.

It was not until the 1980s that particle accelerators were devised that would supply the necessary energies. A group of American scientists at Fermilab in Batavia, Illinois, was striving for it, and a group of European scientists at the European Center for Nuclear Research (CERN) near Geneva, Switzerland, was doing the same at this time.

The task for these two groups was not just a matter of energy. If the particles appeared, they would endure too brief a time to be detected directly. They would have to be identified by their breakdown products (muons and neutrinos), which would have to be detected among large numbers of other particles formed at the same time.

The race between the two laboratories was, therefore, an intricate one. As it happened, Fermilab ran out of money and had less adequate equipment. The CERN was, on the other hand, under the demonically energetic leadership of the Italian physicist Carlo Rubbia (b. 1934), and it won out.

Rubbia modified existing instruments to do the job and, in 1982, obtained 140,000 particle events that might conceivably have resulted in the production of W particles. Making use of computers, these were reduced to but five events that could be explained only as W^- particles in four cases, and W^+ in one. Furthermore, they managed to measure the energy of these particles, and from this worked out their mass, which turned out to be right on the nose, exactly as the theory had predicted.

This was announced on January 25, 1983. Rubbia continued to look for the Z^0 particle, which was 15 percent more massive than the W particles, and therefore harder to detect. In May 1983 it was detected, and the announcement made in June. In 1984, Rubbia received a Nobel prize for this work.

There is one other particle that might exist in connection with the theory. This is the Higgs particle, after the British physicist Peter Higgs, who first proposed its existence. The theory is not clear on what its mass and other properties are. The thought is that it is considerably more massive than the W particles; therefore, no one is really sure when it can be detected. It remains an achievement for the future.

The Electroweak Interaction

We have examined four interactions: strong, electromagnetic, weak, and gravitational, in order of declining intensity. Are there any more? Scientists, generally, are of the strong opinion that there are no more.

But it might be difficult to be sure of this. After all, as recently as 1930, scientists knew of only two interactions, electromagnetic and gravitational, and then the two nuclear interactions turned up. However, the new interactions were not unexpected. The very existence of radioactivity was unsettling in this respect, in that it was clear that neither gravitation nor electromagnetism was useful in explaining it. Once the structure of the atomic nucleus was worked out, there was a shouting need for something new.

The situation today is quite different. In the half-century since the two nuclear interactions were discovered, there has been intense research into every facet of subatomic physics with instruments of unprecedented power and subtlety. In addition, scientists have studied the larger world about us and have probed the Universe with instruments and devices undreamed of in the 1930s.

Discoveries have been made in vast numbers that no one could have predicted, and it is clear that more has been done in the way of scientific investigations and findings in the last fifty years than in all of the thousands of years before.

And yet all of the scientific investigations of the last half century have not isolated a single phenomenon anywhere in the entire range from the Universe to the neutrino that can't be explained by one of the four interactions. The need of a fifth interaction has never shown up, and it is this that leads scientists to believe that four interactions are all there are.

In the late 1980s, to be sure, there was some talk of a fifth interaction that was even weaker than gravitation, that had a range intermediate between that of the nuclear reactions and the other two, and that varied with the chemical compositions of the material involved. For a while, some interest was stimulated by the announcement, but the properties of the interaction were such a tissue of complexity that to me, at least, it seemed very unlikely from the start. As it happened, it faded out rapidly.

Of course, it remains possible that some aspect of the Universe might yet be discovered that lies well outside our knowledge, and that will come as a complete surprise (such as the discovery of radioactivity in 1896). Such a discovery might make necessary the development of additional interactions useful under conditions that until now we have never had occasion to study, but the chances of this seem small.

On the whole, then, the question of the number of interactions is not "Why aren't there more interactions?" but "Why are there as many as four?" Scientists have the definite feeling that there is a principle of economy, so to speak, in the structure of the Universe; that its workings are as simple as possible; that two tasks are not accomplished by two utterly different pathways if they can be done by a single pathway suitably modified to fit both cases.

Thus, as late as 1870, there seemed to be four different phenomena that could make themselve felt across a vacuum: light, electricity, magnetism, and gravitation. All four seemed quite distinct. Nevertheless, Maxwell, as mentioned earlier in the book, in one of the great scientific insights of all time, prepared a set of equations that gov-

267

erned both electricity and magnetism and showed them to be inextricably related. Moreover, if the electric field and the magnetic field were combined into an electromagnetic field, it turned out that light was a radiation intimately and inextricably related to that field. Maxwell could predict an entire family of light-like radiations, from radio waves to gamma rays, with his new insight—radiations not actually discovered until a quarter century later.

It seemed only natural to try to expand Maxwell's treatment further to include the gravitational interaction. Einstein spent the final third of his life at this task, but failed, as did everyone else. Then, in the 1930s, the situation was complicated by the discovery of the two nuclear interactions, the strong and the weak, with science finding itself with four fields again. But this didn't mean scientists gave up trying to find ways of describing all of the fields with one set of equations (a unified field theory). The lure of showing the Universe to be as simple as possible is too strong to be ignored.

In 1967, the American physicist Steven Weinberg (b. 1933) worked out a set of equations that would cover both electromagnetic and weak interactions. The two seemed so different in nature, and yet Fermi had worked out the theory of weak interactions by using the type of mathematics Dirac had used for electromagnetic interactions, so there must be *some* similarity.

Weinberg came up with a treatment that placed the two interactions under a single umbrella, which showed that for what might be called the electroweak interaction there had to be four exchange particles. One was massless, and was undoubtedly the photon. The other three had mass, a great deal of mass, and were what came to be called the W^+, W^-, and Z^0 particles. (There was also the Higgs particle, but that was less certain.)

At about the same time, the Pakistani-British physicist Abdus Salam (b. 1926) produced an almost identical theory, quite independently. (This is not very surprising. It often happens in science that when scientific information reaches a certain level in some field, some startling advance is crying to be made—the time is ready for it, so to speak—and more than one person responds. The most startling case of this sort of thing took place in 1859, when Charles Robert Darwin and Alfred Russel Wallace, independently and simultaneously, made ready to publish the theory of biological evolution by natural selection.)

The electroweak interaction did not achieve immediate recognition and acclaim. The mathematics was incomplete in certain respects, and it was only a few years later that the Dutch physicist Gerard 't Hooft refined the mathematics appropriately.

If the electroweak interaction existed, there should be neutral currents. In other words, there should be particle interactions involving an exchange particle of the weak interaction that did not involve a shift of electric charge from one particle to another. It was for this neutral current that the Z^0 particle was necessary. In 1973, such neutral currents were actually detected, and suddenly the electroweak theory began to look very good. In 1979, Weinberg and Salam received Nobel prizes for it. The actual detection of the weak interaction exchange particles in 1983 put the cap on the theory.

You might wonder, if there is a single electroweak interaction, why the electromagnetic interaction and weak interaction aspects of this single phenomenon are so widely different. Apparently, this is the result of our living at low temperatures. If the temperature were high enough (far, far higher than exists in our environs today), there would really be only one interaction. As the temperature drops, however, the two aspects separate. They are still a single interaction, but are manifested in two widely different forms.

We can make use of an analogy. Water exists in three

forms: liquid water, ice, and vapor. To people unfamiliar with our world, these would seem like three entirely different substances, unrelated to one another.

Now, suppose the temperature were high enough so that all the water would be in the form of vapor. Water would clearly be a single substance with a single set of properties. But let the temperature drop, and some of the vapor would condense into liquid, and the liquid and vapor would remain in equilibrium. Now there would be, apparently, two different substances, with two widely different sets of properties.

If the temperature dropped still lower, some of the water would freeze and you would have ice, water, and vapor all in equilibrium; all three quite different in appearance and properties; and yet all three still the same substance, fundamentally.

There is the thought, therefore, that when the Universe first came into being, it was at an extraordinarily high temperature, something like ten million trillion trillion trillion degrees, and at that time and under those conditions there was only one interaction. As the temperature dropped (very rapidly, as we measure time today), gravitation split off as an apparently separate interaction, which grew weaker as the temperature continued to drop. Then the strong interaction split off, and finally the weak and the electromagnetic interactions split apart.

This makes it seem, naturally, as though the process might be reversed mathematically and that a single treatment might draw all four interactions under one umbrella. Various plans for unifying the electroweak and strong interactions have been advanced, and many scientists are confident of success in developing such a "grand unified theory." So far, however, all attempts to also include gravitation have failed. The phenomenon remains an intractable problem (of which I will have more to say later).

11

QUARKS

The Hadron Zoo

Let's consider the various subatomic particles as we've now described them. First, there are the leptons, which are subject to the weak interaction, and, if electrically charged, to the electromagnetic interaction, but *not* to the strong interaction. These seem to be fundamental particles that have never been shown to have any internal structure. They include three flavors: the electron and its neutrino, the muon and its neutrino, and the tauon and its neutrino. There are also antiparticles for each of these, which brings us to twelve leptons altogether. Scientists don't expect to find any more.

Second, there are the exchange particles, which mediate the four interactions: the graviton for the gravitational interaction, the photon for the electromagnetic interaction, the W particles for the weak interaction, and the pion as Yukawa's exchange particle for the strong interaction. The graviton and the photon are single particles, but the W particles and the pion exist in positively charged, negatively charged, and neutral varieties. This means that there are eight exchange particles altogether. Scientists don't expect to find any more.

This leaves the particles that are subject to the strong interaction. The longest known of these are the baryons; that is, the proton and the neutron, whose existence, cheek by jowl, in the atomic nucleus was the occasion for the development of the strong-interaction theory. In addition, the pions, which are mesons, are subject to the strong interaction.

The particles subject to the strong interaction, the baryons and the mesons, are lumped together as hadrons, from the Greek word for "thick," or "strong." Hadrons are thus a good opposite for leptons, which, as I explained earlier, is from the Greek word for "weak."

If the proton and neutron and their antiparticles, plus the three pions, were all the hadrons there were, there would be seven, a reasonable number. In that the three pions could be counted among the exchange particles, this would mean that leptons, exchange particles, and hadrons, in both their normal form and their antiform, would come to only twenty-four particles altogether, which scientists could live with under a view of the Universe as simple.

However, as particle accelerators grew larger and more efficient, capable of disposing of more and more energy, physicists found that the energy available coalesced, so to speak, into numerous particles that simply didn't exist except under high-energy conditions. These particles were all extremely unstable, enduring at most a millionth of a second, and for the most part for much shorter periods of time.

The new discoveries included the tauon and its neutrino among the leptons, and the W particles among the exchange particles, with all of the rest of the numerous discoveries among the hadrons.

In 1944, for instance, a new particle was discovered that was identified as a meson. It was called the K meson, or, often, the kaon. It had a mass three and a half times that of a pion, and roughly half that of a proton.

In 1947, the first of a group of particles more massive than a proton or neutron was discovered. These were called hyperons, from a Greek word meaning "beyond," in that their masses were beyond those of the proton and neutron, which, until this time, had been thought to be the most massive particles.

This sort of thing continued, and eventually over a hundred hadrons of different sorts were discovered, which implied the existence of a hundred different antihadrons. Some of them endured for only a few trillionths of a trillionth of a second before breaking down, but they were particles just the same.

Scientists were troubled. Every sign had pointed to a satisfactory simplicity of the Universe, and now the "hadron zoo" had reduced things to an apparently meaningless complexity again. Naturally, attempts were made to find order among all of these hadrons; to find ways of grouping them in a meaningful way. If this could be done, one could deal not with many individuals, but with a few groups.

For instance, as early as 1932, Heisenberg pointed out that if one ignored electric charge, the proton and the neutron could be viewed as a single particle in two different states. It was impossible to describe the difference between the states in ordinary terms, but it sufficed to call one state positive and the other negative.

In 1937, the Hungarian-American physicist Eugene Paul Wigner (b. 1902) proposed that the proton and neutron were analogous to isotopes in the periodic table of the elements, and that the two states might be pictured as spins of some sort, in that two spins would account for the difference in states. He called Heisenberg's states isotopic spin, which is now usually shortened to isospin. In 1938, the Russian physicist N. Kemmer pointed out that the three pions—positive, negative, and neutral—could be treated as the same particle in three different isospin states.

The isospin was important, first, because it did group some of the particles and helped ameliorate the hadron complexity, and second because it was conserved among hadrons. This helps make some sense out of the hadron zoo, because all of these particles do not undergo changes and interactions at random, but must conserve various properties. This limits the number of permissible changes. The greater the number of conserved properties that can be worked out, the greater the limitations and the easier it is to understand what is happening.

For instance, kaons and hyperons last a surprisingly long time. It takes a millionth of a second for kaons to break down, and nearly a billionth of a second for hyperons to break down. The mechanics of their production indicates clearly that they are formed through the mediation of the strong interaction, and therefore ought to break down the same way—in a minute fraction of a trillionth of a second.

But they don't; they last for thousands, even millions of times as long as they ought to, and therefore must break down by way of the weak interaction, which seemed strange. In fact, they came to be called strange particles.

In 1953, the American physicist Murray Gell-Mann (b. 1929) suggested there must be a characteristic possessed by strange particles not possessed by other hadrons. He called the characteristic, naturally enough, "strangeness."

The proton, neutron, and various pions each had a strangeness number of 0, but kaons and hyperons did not. The strangeness number was conserved in the strong interaction. Kaons and hyperons could not break down by way of the strong interaction because they formed pions and protons with strangeness numbers of zero, which meant the disappearance of strangeness and a resultant violation of the law of conservation. Kaons and hyperons must therefore break down by means of the weak interaction, in which strangeness need not be conserved. This is why strange particles endure so long.

Studies of the hadrons did not always succeed in establishing or preserving laws of conservation. In one case, the revision of a law of conservation was forced on scientists.

As early as 1927, Wigner had advanced the law of conservation of parity. Parity cannot be explained literally, but we can deal with an analogy here, in terms of odd and even number. Two even numbers always add up to an even number, and two odd numbers always add up to an even number. However, an even number and an odd number always add up to an odd number. If, then, we call some particles even and some odd, the permissible changes must adhere to the same rules: even + even = odd + odd = even; while even + odd = odd + even = odd.

But then, in the early 1950s, it was found that a particular variety of kaon had a peculiar way of decaying. Sometimes it decayed into two pions and sometimes into three. The two pions added up to even parity, but the three added up to odd parity. The question was, then: How could the kaon be both odd and even?

The easiest way out was to suppose that there were actually two very similar particles, one of which was odd parity and one even parity. These were named tau meson and theta meson after two letters of the Greek alphabet. This would have settled the issue except that there seemed no way to distinguish between the tau meson and the theta meson.

This is not a deadly situation, however. The muon neutrino cannot be distinguished from the electron neutrino by any measurable property, but only by their behavior in various interactions. Perhaps this was true of the tau meson and theta meson, too.

In the case of the two neutrinos, however, there had seemed no alternative but to accept an indistinguishable difference. In the case of the two mesons, there was. What if parity was not always conserved?

The Chinese-American physicists Chen Ning Yang (b. 1922) and Tsung-Dao Lee (b. 1926) worked out the theoretical consequences of this in 1956, convinced that parity was *not* conserved, at least in those reactions mediated by the weak interaction. But how could this be demonstrated?

The answer lay in the fact that, in a way, the conservation of parity was equivalent to the notion of left-right symmetry. In other words, if parity is conserved and if a certain interaction produces a stream of particles, these particles will come off to the left and to the right in equal numbers. If, however, parity is not conserved, the particles will come off only to the left, or only to the right. (One of the reasons why scientists found it so difficult to believe that parity was not conserved was that they saw no reason why the Universe should distinguish between left and right.)

An experiment was, therefore, arranged at Columbia University, with another Chinese-American physicist, Chien Shung Wu, in charge. She worked with a sample of the radioactive isotope cobalt-60, which broke down to yield beta particles, mediated, of course, by the weak interaction. These beta particles came off in all directions, partly because the atoms themselves faced in all directions. Wu therefore placed the material in a strong magnetic field so that all of the atoms would line up in the same direction. That would give them a chance to fire off the beta rays in one direction only if parity was not conserved. Of course, at ordinary temperatures, the atoms would wriggle themselves into different directions despite the constraint of the magnetic field; therefore, Wu cooled the cobalt-60 very nearly to absolute zero.

If parity was not conserved, the beta particles should come off only on one side. By January 1957, there was no doubt: the beta particles were coming off in only one direction, and parity was not conserved in weak interactions. That year Yang and Lee received a Nobel prize.

Parity was conserved in other types of interaction and, even in the weak interaction, a more general law of conservation could be substituted. If a particular particle was "left-handed," in terms of parity (P), its antiparticle with an opposite charge (C) was right-handed. This meant that if a particle and its antiparticle were taken together, the property of CP (parity and charge both taken into consideration) would be conserved.

But then, in 1964, the American physicists Val Logsden Fitch (b. 1923) and James Watson Cronin (b. 1931) showed that even CP wasn't always conserved. The property of time (T) had to be added. If CP wasn't conserved in one direction of time, it wasn't conserved, in the opposite way, in the other direction. It is now believed that CPT symmetry is what is conserved in weak interactions. Cronin and Fitch shared the Nobel prize in 1980 for this work.

In 1981, Gell-Mann began using a number of conserved characteristics to group hadrons in symmetrically formed polygons containing eight, nine, or ten individuals. He thus set up families of particles and introduced something analogous to the periodic table of elements. At the same time, the Israeli physicist Yuval Ne'eman (b. 1925) was doing the same thing.

It was difficult for scientists to take Gell-Mann's arrangements seriously, just as it had been difficult for them, a century earlier, to take Mendeleev's periodic table seriously. Mendeleev, however, had won them over when he used the table to predict the properties of undiscovered elements—and had proved to be right.

Gell-Mann envisioned a triangle of ten particles, arranged so that the values of different conserved properties varied in a fixed and regular way from point to point. However, the point at the apex did not correspond to any particle known at the time.

From the arrangement, it could be seen that the missing particle had peculiar properties, including an unusually high mass and an unusually high strangeness number. It was called the omega minus particle, whose existence had to be taken with some skepticism.

From the nature of the properties of the omega minus particle, Gell-Mann believed it must be produced by the interaction of a negative kaon with a proton. These would have to be smashed together at energies high enough to produce a particle with the unusually large mass attributed to the omega minus.

Gell-Mann then had to persuade someone in control of a large particle accelerator to try the experiment. In December 1963, the team at the accelerator at Brookhaven, Long Island, began smashing K mesons into protons. On January 31, 1964, an event was detected that could only have involved an omega minus particle, for a particle showed precisely the properties predicted by Gell-Mann. In 1969, Gell-Mann received a Nobel prize for his work. Now Gell-Mann's groupings of hadrons had to be taken seriously. The hadron zoo was yielding to order.

Inside Hadrons

Merely dividing hadrons into groups and setting up a type of subatomic periodic table wasn't enough. Mendeleev's periodic table wasn't satisfactorily explained until the internal structure of atoms was worked out and the significance of the difference in electron arrangements within their shells was understood.

It seemed to Gell-Mann, then, that there had to be an internal structure in regard to hadrons that would account for their existence in groups. This was by no means an untenable idea. The leptons were fundamental particles that behaved as though they were simple points in space without internal structure, but this was not necessarily true of the hadrons.

What Gell-Mann tried to do, then, was to make up a group of particles that would perhaps be fundamental, with properties such that if they were put together properly they would form all of the various hadrons with *their* properties. One combination would yield the proton, another the neutron, still another the various pions, and so on.

Gell-Mann set about the task and quickly found that he could not manage to do this if he stuck to the principle that every particle must have electric charges either equal in size to the familiar charge on the electron or proton or a multiple thereof. He found, instead, that the constituent particles of the hadrons would have to have fractional charges.

At this, Gell-Mann quailed. In all of the time people had been working with electrically charged particles, dating back to the very beginning of Faraday's researches on electrochemistry a century and a third earlier, charges had seemed to come in even multiples, the smallest (and apparently indivisible) of which had for three quarters of a century been considered that on the electron.

In 1963, however, Gell-Mann decided to publish anyway. He suggested that there would be three fundamental particles making up the hadrons, and three antiparticles making up the antihadrons. Each hadron was made up of

279

either two or three of these fundamental particles. The mesons were made of two, and the baryons of three.

Gell-Mann called these fundamental particles quarks. (It might have been intended as a bit of whimsy taken from James Joyce's *Finnegans Wake,* in which appears the phrase, "Three quarks for Muster Mark." This I have always taken to mean, in Joycean language, "Three quarts for Mister Mark," and supposed it to represent an order for beer. To Gell-Mann, it seemed there were "three quarks for Muster Hadron." The name should not have been kept, in my opinion. It is inelegant. However, the name stuck—possibly to Gell-Mann's own surprise—and is now ineradicable.)

Gell-Mann specified three types of quarks, which were whimsically called the up quark, the down quark, and the strange quark. (The adjectives are not to be taken literally, of course. One can speak of the u quark, the d quark, and the s quark, or symbolize them simply as u, d and s. The s is sometimes said to stand for "sideways" quark to have it harmonize with *up* and *down,* but *strange* is better because it is more significant.)

The u quark has an electric charge of $+\frac{2}{3}$, and the d quark of $-\frac{1}{3}$. (Naturally, the u antiquark has a charge of $-\frac{2}{3}$, and the d antiquark of $+\frac{1}{3}$.) Each type of quark has a series of numbers representing the various characteristics it conserves. The quarks have to be put together in such a way that the hadron they form has all of the various proper numbers for its characteristics.

Naturally, it is the fractional charges that one must be most careful of. Quarks also have to be put together in such a way that the total electric charge on the hadron comes out to be $+1$, -1, or 0. For instance, a proton is built of two u quarks and one d quark; therefore, its total charge is $+\frac{2}{3}$ and $+\frac{2}{3}$ and $-\frac{1}{3}$, or $+1$. An antiproton is built of two u antiquarks and one d antiquark ($-\frac{2}{3}$ and $-\frac{2}{3}$ and

+ ⅓), for a total charge of −1. A neutron is built of one u quark and two d quarks (+⅔ and −⅓ and −⅓), for a total charge of 0, and an antineutron is built of one u antiquark and two d antiquarks (−⅔ and +⅓ and +⅓), for a total charge of 0.

A positive pion is built of a u quark and a d antiquark (+⅔ and +⅓), for a total charge of +1, and a negative pion is built of a u antiquark and a d quark (−⅔ and −⅓), for a total charge of −1.

The s quark goes into the making of the strange particles, which is where it got the s of its name. The s quark has an electric charge of −⅓ and a strangeness number of −1. The s antiquark has an electric charge of +⅓ and a strangeness number of +1.

The positive K meson contains a u quark and an s antiquark (+⅔ and +⅓), for a total charge of +1 and a strangeness number of +1. The negative K meson consists of a u antiquark and an s quark (−⅔ and −⅓), for a total charge of −1 and a strangeness number of −1.

A lambda particle (a neutral hyperon) consists of a u quark, a d quark, and an s quark (+⅔ and −⅓ and −⅓), for a total electric charge of 0, whereas an omega minus is made up of three s quarks (−⅓ and −⅓ and −⅓), for an electric charge of −1. Both the lambda and the omega minus are strange particles.

In this sort of way, the various hadrons are built up, and no combination is possible that doesn't yield a total charge of either 0, +1, or −1.

But is all of this really true? Do the quarks really exist, or is this just bookkeeping? After all, a dollar bill is worth any of various combinations of coins—half-dollars, quarters, dimes, nickels, and pennies—but if the dollar bill is torn to bits it turns out there are no coins making up any part of its structure.

Suppose, then, you pull a hadron apart; will quarks

come tumbling out? Or is *this* just bookkeeping? Unfortunately, no one has yet succeeded in tearing a hadron apart or in unequivocally producing a free quark. If one were produced, it would be easy to identify because of its fractional charge. However, there are some scientists who think that it is impossible, even in theory, to pull a quark out of a hadron. And even if it were possible, we certainly don't yet dispose of energies high enough to turn the trick. However, there is indirect evidence for the actual existence of quarks.

In 1911, Rutherford described the experiments he had done in bombarding atoms with alpha particles. The alpha particles, for the most part, passed through the atoms as though they were nothing but empty space, but there was some scattering. Every once in a while they hit some small objects within the atoms and were deflected. From this Rutherford deduced that there was a minute massive point within the atom—the atomic nucleus.

Might it not be possible to bombard protons with very high-energy electrons and thereby make them scatter? From the results, might it not be possible to deduce that there were scattering points within the proton, and therefore that quarks really existed in there?

Such experiments were carried out at the Stanford (University) Linear Accelerator in the early 1970s by Jerome Friedman, Henry Kendall, and Richard Taylor, who received the Nobel prize in physics in 1990 as a consequence. The results were satisfactorily interpreted by the American physicist Richard Phillips Feynman (1918–1988), who had already received a Nobel prize in 1965 for something I'll mention later. By 1974, it was clear that quarks really *did* exist, even if they were never spotted in the free state.

Feynman referred to these particles inside the protons as partons. (This, to my way of thinking, is a much better name than quarks. Either Feynman thought, as I do, that quark was uneuphonious, or he thought that Gell-Mann's quark theory was not quite the way it ought to be.)

But now comes the possibility of trouble. We got down to atoms and there proved to be so many different types of them that simplicity was lost. We went down a notch to subatomic particles to restore the simplicity, and there proved to be so many different types of these that simplicity was lost a second time. Now that we are down to quarks, will it turn out that there are a great many different types of them?

Some people thought there ought to be at least one more quark. One among them was the American physicist Julian Seymour Schwinger (b. 1918), who had shared a Nobel prize with Feynman in 1965. It seemed to Schwinger that the quarks were fundamental particles like the leptons. He believed that they were point particles without internal structure (whose diameter is zero as nearly as we can determine) and that there ought to be symmetry between these two sets of fundamental particles.

Two flavors of leptons were known at the time—the electron and its neutrino, and the muon and its neutrino— and thus two flavors of antileptons. There should thus be two flavors of quarks. One flavor was the u quark and the d quark (and its antiquarks, of course). The second flavor was the s quark and—what? If a fourth quark existed, particles containing it had not been found, but that might be because the fourth quark and the particles containing it were so massive that considerable energy was needed to produce it.

In 1974, a team led by the American physicist Burton Richter (b. 1931) made use of the powerful Stanford (University) Positron-Electron Accelerating Ring, producing a particle that was massive indeed—three times as massive as a proton. A particle that size should break down in the merest evanescence of a second, but it didn't, it hung on. Therefore, it had to contain a new quark—one that, like the s quark (but much more massive than the s quark), prevented breakdown by way of the strong interaction.

The new particle was called a charmed particle because it lived so long and, presumably, contained a "charmed quark" or c quark, which was the fourth quark that Schwinger had been looking for. It was, indeed, more massive than the other three. The same work was done, and the same conclusions reached, at the same time by the American physicist Samuel Chao Chung Ting (b. 1936) at Brookhaven. Richter and Ting shared a Nobel prize in 1976.

By this time, however, a third flavor of leptons had been discovered, in the form of the tauon and the tauon neutrino (and its antiparticles). Did this mean there should be a third flavor of quarks?

In 1978, a fifth particle was indeed discovered, which was called the bottom quark, or b quark. There must be a sixth, which scientists call the top quark or t quark, but it hasn't been located yet, presumably because it is excessively massive. (Some scientists prefer to have the b and t stand for "beauty" and "truth.")

Quantum Chromodynamics

We now have three flavors of quarks, as we have three flavors of leptons. In each flavor there are two leptons, or quarks, and two antileptons and antiquarks. This means that there are 12 leptons and 12 quarks altogether. There are 24 particles that, with the exchange particles, make up the entire Universe (or so it now appears). This puts us back to a tolerable simplicity—at least for now. As I shall explain later, the situation might not last.

The similarities between the two types of particles are

interesting. In the case of leptons, the first flavor consists of an electron with a charge of -1 and an electron neutrino with a charge of 0. This pattern is repeated in the other two flavors: a muon with a charge of -1 and a muon-neutrino with a charge of 0, and a tauon with a charge of -1 and a tauon-neutrino with a charge of 0. Naturally, this is reversed for the antileptons, where all three flavors have charges of $+1$ and 0.

In the case of quarks, the first flavor includes the u quark $(+\frac{2}{3})$ and the d quark (-1.3). This pattern is repeated in the second and third flavors with the c quark $(+\frac{2}{3})$ and the s quark $(-\frac{1}{3})$ and with the t quark $(+\frac{2}{3})$ and the b quark $(-\frac{1}{3})$. Again, this pattern is reversed in the case of antiquarks.

Of course, the comparison isn't exact. The leptons include particles with integral charge and zero charge, while quarks have neither, including only particles with fractional charge.

Again, the masses of the particles goes up with the flavor in the case of the charged leptons (the uncharged neutrinos are massless). If we set the electron's mass at 1, the mass of the muon is 207 and that of the tauon is about 3,500. The mass goes up with flavor in the case of the quarks, also, but there are no massless quarks, perhaps because there are no uncharged quarks.

In the case of the first flavor of quarks (if we still consider the electron's mass 1), the u quark, which is the least massive of all of the quarks, has a mass of 5, and the d quark has a mass of 7. In the case of the second flavor, the s quark has a mass of about 150 and the c quark has one of about 1,500. The c quark is almost as massive as a proton, which is why it takes so much energy to produce charmed particles and why their discovery came so late.

The third flavor is more massive still. The b quark has a mass of about 5,000, or nearly three times the mass of a proton, which is why it was discovered even later than the c quark. The t quark, not having been located, has no reliable figure for its mass, but the estimate is that it must be up to at least 25 times the mass of a proton, which is why it hasn't been discovered.

It is not enough, of course, simply to list all of the quarks. One has to make sense out of the mechanism by which they work. In 1947, for instance, three physicists worked out, independently, three somewhat different ways of describing exactly what happens in the interaction of electrons and photons, thus explaining the mechanism of the electromagnetic interaction. All three ways were valid, and were essentially equivalent.

Two of the physicists were Schwinger and Feynman. The third was the Japanese Sin-itiro Tomonaga (1906–1979). (It might be that Tomonaga had it first, but World War II was raging and Japanese scientists were isolated. Tomonaga could not publicize his ideas until after the war.) All three shared a Nobel prize in 1965.

The theory is called quantum electrodynamics, which turned out to be one of the most successful theories ever devised. It predicted events involving the electromagnetic interaction with phenomenal accuracy, and has not been improved on since its formulation.

Naturally, scientists thought that the techniques used in quantum electrodynamics could be used to work out the details of the strong and weak interactions, but in this they were at first disappointed. Finally, Weinberg and Salam were able to unify the electromagnetic and weak interactions, but the strong interaction continued to present problems.

For instance, quarks have half spins and are therefore fermions, just as the leptons are. There is an exclusion principle first worked out by Pauli in 1925 that states that two fermions cannot be grouped into the same system if all of their quantum characteristics are identical. There always has to be some difference in the quantum numbers assigned them. If an attempt is made to squeeze together two fermions with identical quantum numbers, there is a repulsion between them far larger than electromagnetic repulsion. Nevertheless, it turns out that in some hadrons three identical quarks can be squeezed into one hadron, just as though the exclusion principle did not exist. The omega minus particle, for instance, is built of three s quarks.

There was, however, great reluctance to give up the exclusion principle, which worked everywhere else in subatomic physics, and scientists were anxious to save it in the case of quarks. It might be that there was some distinction among quarks that were otherwise apparently the same. If there were, for instance, three varieties of s quarks, one of each variety might be squeezed into a hadron without violating the exclusion principle.

Beginning in 1964, several physicists—among them Oscar Greenberg at the University of Maryland, Japanese-American physicist Yoichiro Nambu (b. 1921) at the University of Chicago, and Moo-Young Han at Syracuse University—worked on this matter of quark varieties.

They decided that the varieties were not something analogous to anything else in subatomic physics, and could not really be described. It could only be given a name, and the manner in which it worked detailed. The name given the distinction was *color*.

In a way, of course, this is a bad name, for quarks do not have color in the everyday sense. In another way, though, it is perfect. In color photography and color television, it is well known that the colors red, green, and blue will combine to give the impression of colorlessness; that is, whiteness. If every quark comes in red, green, and blue varieties, a combination of one of each leads to a disappearance of color, to whiteness. Every quark combination in hadrons must produce a white result. No hadron is known in which there is a color because the quark content is color imbalanced.

This explains why there are three quarks to every baryon, and two quarks (or, rather, a quark and an antiquark) in every meson. These are the only combinations that are colorless.

Once color was taken into account, several observations that had been anomalous without it could be seen to be right on the nose with it. For this reason, the notion of colored quarks was quickly adopted by scientists.

Of course, if there are six different quarks and six different antiquarks among the three flavors, and if each quark comes in three different colors, then there are thirty-six colored quarks altogether. This increases the complexity of the situation, but it gives scientists a handle with which to evolve a theory of quark behavior that approximates the value of quantum electrodynamics. The new theory is called quantum chromodynamics, where chromo- is from the Greek word for "color." Much of this new theory was worked out in the 1970s by Gell-Mann, who had first suggested the quark concept.

The strong interaction is essentially that between quarks. Hadrons, which are made up of quarks, experience the strong force secondarily *because* they are made up of quarks. Pions, which seem to be the exchange particle for this secondary hadron interaction, are exchange particles only because they, too, are made up of quarks. In other words, all of the emphasis on the *fundamental* strong interaction must be shifted to the quarks.

If this is so, there must be some exchange particle that exists on the quark level. It was Gell-Mann who came through with a name for this new exchange particle. He called it a gluon because it was the glue that held quarks together.

Gluons had properties that were quite unusual. Thus, in the case of the other exchange particles, the greater the distance between particles subject to the interaction, the

If you try to pull two quarks apart, the number of gluons between them increases. This is the attractive force between quarks, which actually *increases* with distance. *Thus quarks can only move about freely within the hadron. Scientists suspect that they may never be able to study free quarks.*

fewer the exchange particles bouncing between them and the weaker the interaction. The gravitational and electro-magnetic interactions grow weaker at a value the square of the distance between the two objects subject to these interactions. The weak interactions, and the secondary strong interaction between hadrons, decline in intensity even more rapidly with distance.

In the case of quarks and gluons, however, it is quite the other way around. If you try to pull two quarks apart, the number of gluons bouncing between them *increases*. This is equivalent to saying that the attractive force between quarks increases with distance.

Within the hadron, the quarks move about freely and easily. They stiffen, however, as they move apart. This means that quarks undergo particle confinement; they can only exist comfortably inside the hadrons. It is for this reason that scientists suspect we are not ever going to be able to study free quarks. There's no way of making them leave the hadrons. Of course, the hadrons themselves can change from one to another, carrying their load of two or three quarks (which may themselves change from one color to another) within themselves.

Gluons are more complex than other exchange particles in another way. Gravitons are exchanged by particles with mass, but gravitons themselves have no mass. Photons are exchanged by particles with electric charge, but photons themselves have no electric charge. Gluons, which themselves have color, are exchanged by particles with color; therefore, gluons can stick to each other. This is another reason why the name gluon is a good one. (Some scientists refer to a pair of gluons that stick together as a glue ball.)

The gluon has the capacity of changing the color of a quark (but not the flavor). There is a gluon that changes a red quark to a green quark of the same flavor, another that changes red to blue, and so on. To account for all of the color changes possible, there must be eight different gluons. This is an added complexity. With one gravitational exchange particle, one electromagnetic, three weak, and now eight strong, we have thirteen exchange particles altogether.

Nevertheless, quantum chromodynamics based on quarks in three flavors and three colors, with eight colored gluons (forty-four particles altogether), is a successful theory, and scientists expect that it will continue to explain all of the ins and outs of hadrons and their behavior.

12

THE UNIVERSE

The Mystery of the Missing Mass

Naturally, the observations and experiments scientists have made in connection with subatomic physics are, for the most part, conducted right here on Earth. How do we know that the results we get are applicable to other worlds—to the stars or to the Universe in general?

To begin with, we have made studies of the surface of the Moon, of Mars, and of Venus directly, and we have studied the surface of other objects in our solar system by rocket probes employing sophisticated instruments for the purpose—even if they have not made actual physical contact. We even have bits of extraterrestrial matter that arrive on Earth in the form of meteorites. None of these investigations have offered us any subatomic surprises. Scientists are quite certain that all of the planetary objects in the solar system are made up of the same matter that Earth is, and therefore must follow the same rules.

But what about the Sun, which seems so different from all of the other members of the solar system? Well, charged particles, mostly protons, reach us from the Sun, as do neutrinos, and they are what we expect them to be.

What about the Universe beyond the solar system? We have received neutrinos from the supernova that exploded in 1987 in the Large Magellanic Cloud, and we receive cosmic rays (mostly protons and alpha particles) from the Universe generally. They indicate that the Universe follows the rules that have been worked out here on Earth.

The most important information that reaches us from the Universe generally comes in the form of photons. We actually *see* the Sun and the stars, and we even see galaxies that are billions of light-years away. We can also detect photons that are too energetic, or unenergetic, for our eyes to see—gamma rays, X rays, ultraviolet, infrared, and radio waves.

The photons we get give a clear indication of the chemical structure of the objects that emit them. Astronomers are quite satisfied that the stars and galaxies are made up of matter like that that makes up our own Sun. Our Sun is made up of matter like that on Earth (allowing for the Sun's much higher temperature).

But do we really see, or sense, a fair sampling of all of the photons there are? Is there anything that exists in the Universe that *doesn't* radiate photons? Not really! Every object in the Universe that is surrounded by space at the average temperature of the Universe (about three degrees above absolute zero)—and that means just about every object—radiates photons. Some of the radiation, however, is either insufficiently intense or insufficiently energetic for us to pick up.

There are many stars that are so dim that unless they are fairly close to us we cannot see them, even with the best instruments we now have available. There are certainly planets in other star systems that are as surface-cold as the planets of our own solar system, and whose feeble radiation of radio waves is lost in the blaze of the stars they circle.

Nevertheless, it seemed fair to assume that by far the largest percentage of the mass of the Universe took the form of stars, and that the amount of mass we couldn't sense, because it was too cold and faint, was not significant. In our own solar system, for instance, all of the planets, satellites, asteroids, comets, meteors, and dust that circle the Sun make up only 0.1 percent of the total mass. The other 99.9 percent of the mass is found in the Sun. There would seem every reason to suppose that, by and large, other stars also predominated in this way over the objects circling them.

Naturally, it is possible that there are places in the Universe in which conditions are so extreme that the laws of nature we have worked out don't necessarily hold. The most likely place where this might be true are in black holes, where matter has collapsed into conditions of nearly infinite density, creating a small neighborhood of nearly infinite gravitational intensity. We can't study black holes in detail and, as yet, we have not even completely and unmistakably identified any. However, assuming they exist, they might be governed by laws outside those we know.

Another realm of uncertainty exists in the first instants after the Universe came into existence, when conditions were so extreme that our structure of physical theory might not apply. (I will have a few words to say about this later.) Yet nothing ever seems to be surprise-free. All of the photons from the outside universe that we study are the product of the electromagnetic interaction, and the surprise arose from the effects of gravitation, the other long-range interaction.

We can't detect gravitons, but we *can* detect the effect of gravitation on the movement of stars and galaxies. We can measure the speed of rotating galaxies in different parts of their structures, and we assume that this rotation is driven by gravitational forces within the galaxy, just as the rotation of the planets of the solar system is driven by the gravitational influence of the Sun.

Because 99.9 percent of the mass of the solar system is concentrated in the Sun, the solar gravitational influence overwhelms everything else in the solar system. Except for very minor corrections, this influence alone needs to be taken into account. The farther a planet is from the Sun, the less intense is the Sun's gravitational influence on it, and the more slowly it moves. The variation in motion with distance was first worked out in 1609 by the German astronomer Johannes Kepler (1571–1630), and was explained by the law of universal gravitation, advanced in 1687 by Newton.

Like the solar system, galaxies have their mass concentrated at the center, although not to quite such an extreme. We can *see* that the stars are more and more numerous as one approaches the center of a galaxy, and it seems a fair conclusion that about 90 percent of the mass of all large galaxies is contained in a relatively small volume at their cores. Therefore, we would expect the stars to be circling the center of their galaxies more and more slowly as one moved outward from the core. But this does not happen. Apparently, the stars in a galaxy move at about the same speed as one moves outward from the core.

No scientist wants to abandon the law of gravity (which has been modified and extended, but not replaced, by Einstein's theory of general relativity), in that it would seem that no alternative law can explain what goes on in the Universe generally. Therefore, we must suppose that the mass of the galaxy is *not* concentrated at the core, but is spread out much more evenly throughout the galaxy. Yet how can this be when we *see* that the mass, in the form of stars, *is* concentrated?

The only conclusion we can come to is that there is matter outside the core that we *don't* see. It is "dark matter" that doesn't send us anything in the way of perceptible photons, but that exerts its gravitational influence. In fact, we are forced to assume, from a gravitational standpoint, that the mass of a galaxy might be many times as great as it would seem to be from the photons it radiates. Until studies of galactic rotations were made, we were apparently missing most of the mass of the galaxies.

Another point: galaxies exist in clusters. Within the clusters (each of which is made up of anywhere from dozens to thousands of galaxies), the individual galaxies move about restlessly, like a swarm of bees. The clusters are held together by the mutual gravitational attraction of the galaxies that make them up, but the masses of the galaxies— if we go by just what we can *see*, by the photons we can detect—are simply not great enough to supply the necessary gravitational pull to hold the clusters together. Yet the clusters apparently *do* hold together. Again, there must be mass that we're not aware of. The larger the cluster, the larger the quantity of mass we cannot quantify. There might be as much as 100 times the mass in the Universe as that we can see. This phenomenon is what is called the "mystery of the missing mass." What is it?

The easiest answer is to suppose that every galaxy contains vast crowds of small, very dim stars, planets, and dust clouds. The trouble is that it isn't reasonable, from what we know about the Universe, to suppose that such material is likely to exist in such quantities that its mass would be a hundred times as great as that of the stars we can see.

Let's get down to the subatomic world, then. About 90 percent of the mass of the Universe, as far as we know, is made up of protons. The only other subatomic particles to match or exceed the number of protons are the electrons, which are equal to the protons in number, and the photons and electron neutrinos, each of which might, in numbers, be a billion times as many as the protons. However, the electrons have only minute masses, and the photons and electron neutrinos have no intrinsic mass at all. The electrons, photons, and electron neutrinos are all speeding along and have energies of motion equivalent to their masses, but the masses that produce the energy are extremely small—so small as to be neglected. This leaves only the proton as the mass material of the Universe.

Is it conceivable that the missing mass is made up of additional protons that we are not aware of? The answer to this seems to be no! Astronomers have ways of estimating the density of protons in the Universe, and therefore of determining how many there can be, seen or unseen, in the regions taken up by galaxies or by clusters of galaxies. The amount of protons present is, at most, only 1 percent of the missing mass. Whatever the missing mass is, then, it can't be protons.

This leaves the electrons, photons, and electron neutrinos. We are quite certain that the electrons and photons can't possibly supply the missing mass, but we are a little less certain about the electron neutrino.

In 1963, a group of Japanese scientists suggested that the electron neutrino might have a minute mass; just a small fraction of that of electrons, for instance. If this is so, the muon neutrino might have a slightly larger mass, and the tauon neutrino a still larger one. All of the masses might be very small, but not quite zero.

If this were so, the neutrinos would travel at less than the speed of light—though perhaps not much less—and each would travel at a slightly different speed. Therefore, the three neutrino flavors would oscillate, shifting from one flavor to another rapidly.

This would mean that if there were a beam of electron neutrinos starting from the Sun, some eight minutes later, when it had completed its 150-million-kilometer race to the Earth, it would appear on Earth as a beam of equal quantities of electron neutrinos, muon neutrinos, and tauon neutrinos.

This would be interesting indeed, in that Reines, who has been detecting neutrinos from the Sun for decades, uses detecting devices that work only for electron neutrinos. If the neutrinos are oscillating, he would be receiving a beam constituted of only one-third electron neutrinos, instead of entirely electron neutrinos. He would detect only one-third, which would explain why the electron neutrino count he received was always so low.

In 1980, Reines reported that he had conducted experiments that gave him reason to believe that oscillation was taking place and that neutrinos *did* have a very small mass. If so, this would explain not only the missing neutrinos from the Sun, but the mystery of the missing mass. There are so many neutrinos floating around the Universe that even if each one had a mass of only $\frac{1}{10,000}$ that of an electron, this would be enough to make the total mass of neutrinos a hundred times that of the mass of all of the protons in the Universe. Moreover, such slightly massive neutrinos might be used to explain how the galaxies formed in the first place, a problem that is giving astronomers a great deal of headache material right now.

The possibility of a slightly massive neutrino would thus very nearly solve a number of problems, and it makes one ache to believe that such a situation is true. The only trouble is that no one has confirmed Reines's report. In

297

general, it is thought he was wrong. No matter how beautiful and desirable a theory is, if it doesn't match the Universe, it must be given up.

But even if the missing mass is not protons and not neutrinos, it still seems to exist. What is it, then? Physicists have, in recent years, been trying to work out theories that unify the strong interaction and the electroweak interaction. Some of these theories require the invention of new and exotic particles. Perhaps it is such particles, never actually observed, and existing, so far, only in the minds of some imaginative scientists, that account for the missing mass. If so, we must wait for observations that will back up these far-out theories.

The End of the Universe

To the casual observer, the Universe, whether seen by eye or by various instruments, might seem unchanging. What changes do take place are likely to be cyclic. If some stars explode, others are formed. There would seem to be no reason to think that the Universe necessarily has an end or a beginning, except for one overwhelming effect, which might not be cyclic: The Universe is expanding.

This story began in 1912, when the American astronomer Vesto Melvin Slipher (1875–1969) began studying the spectra of certain nebulas. These were actually distant galaxies lying far outside our own Milky Way, but that was not understood at the time. From the spectra, Slipher could tell whether the spectral lines were shifted toward the violet end of the spectrum (in which case the nebula was approaching us), or toward the red end (in which case it was receding from us).

By 1917, Slipher found that of the fifteen nebulas he

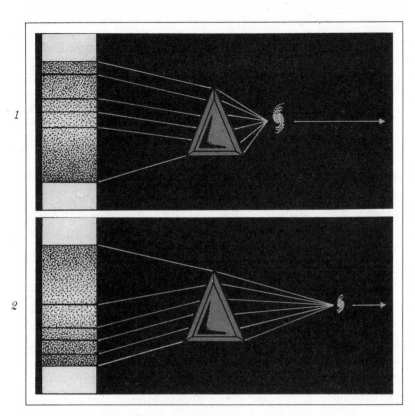

1. *Spectral lines in the light arriving from a nearby galaxy (moving away) will only be slightly shifted toward the red end of the color spectrum.*

2. *Spectral lines from the light of a distant galaxy (moving rapidly away) will be greatly shifted toward the red end of the color spectrum.*

had studied, all but two showed a red shift and were receding. Others took up the task, and when the nebulas were recognized as distant galaxies, it turned out that, barring the two approaches that Slipher had noted (which were two galaxies unusually close to us), all were receding. Moreover, the dimmer the galaxy, the more rapidly it was receding.

By the end of the 1920s, the American astronomer Edwin Powell Hubble (1889–1953) had collected enough data to be able to announce that the Universe was expanding, and that the clusters of galaxies that made up the Universe were receding from one another.

This made sense in the light of theory. In 1916, Einstein evolved his theory of general relativity, which described gravitation more accurately than Newton had done. The equations Einstein worked out to describe gravity were, in effect, the founding of the science of cosmology (the study of the Universe as a whole).

Einstein assumed at first that the Universe would have to be unchanging overall, and adjusted his equations to fit this assumption. In 1917, the Dutch astronomer Willem de Sitter (1872–1934) showed that the unadjusted equations, if properly solved, implied that the Universe was expanding. Hubble's observations proved this theory to be correct.

Now the question is: How long will the Universe continue to expand? Resisting expansion is the mutual gravitational attraction to one another of all parts of the Universe. The expansion, then, is taking place against the pull of gravity, just as an object hurled upward from the surface of the Earth moves against the pull of Earth's gravity.

It is our common experience that an object sent upward under ordinary circumstances is eventually defeated by the gravitational pull of Earth. Its speed of ascent decreases to zero, whereupon the object begins to be pulled back to Earth. The more forcefully the object is hurled upward— and therefore the greater its initial upward speed—the higher it climbs and the longer it takes to begin to fall back.

If an object is sent up from the Earth with sufficient force (sufficient initial speed), it *never* falls back. The gravitational pull of the Earth weakens as the object places more and more distance between itself and Earth's center. If the object moves upward rapidly enough (11 kilometers per second, or 7 miles per second), the declining gravitational intensity never suffices to bring it back. This means that 11 kilometers per second is the escape velocity from Earth's gravitational pull.

We might ask, then, whether or not the rate of the Universe's expansion outward against gravitation's inward pull had achieved escape velocity. If its expansion speed is over escape velocity, the Universe will expand outward forever. It would then be an open Universe. If, however, its expansion speed is below escape velocity, the expansion will gradually slow, and eventually come to a halt. After this, the Universe will begin to contract. It would then be a closed Universe.

It can't possibly affect us in our individual lifetimes, or even in the lifetime of our planetary system, whether the Universe is open or closed, whether it will end as an ever-expanding and thinning ball of matter, or whether it will end as a contracting and thickening ball of matter, but scientists are curious. In order to come to a decision, they try to estimate the rate of expansion. They also try to estimate the average density of matter in the Universe, which gives them an idea of the strength of the inward gravitational pull. Both determinations are difficult to carry through, and the results are only approximate. The conclusion is, however, that the density of the Universe is only about 1 percent or so of that required to end the expansion. The Universe would therefore seem to be open and to be expanding forever.

But wait! The determination of the density of the matter in the Universe is based on what we can detect—but what about dark matter? If it is true that the dark matter in the Universe, the matter whose nature we have not yet determined, is perhaps as much as a hundred times as massive as the matter we *can* detect, this might be enough to close the Universe. We end, therefore, by being uncertain as to whether the Universe is open or closed.

It is also possible that there is enough dark matter in the Universe to place it just on the boundary (or very near it) of open and closed, meaning that the Universe is "flat." This would be an extraordinary coincidence, and the feeling is that if the Universe is flat, there must be a reason for it.

You see, then, how important it is, from a cosmological standpoint, that we know whether or not dark matter is really there, and, if it is, just what it consists of. The answer, when it comes, is bound to arise out of the realm of subatomic particles. Thus, we see that the advance of knowledge is truly unitary. Knowledge of the greatest object we recognize, the Universe, depends on what we know of the smallest objects we recognize, the subatomic particles.

Another way in which subatomic particles might effect the end of the Universe arises out of the attempted unification of the strong and electroweak interactions. The first attempts in this direction began in 1973, when Salam, the cofounder of the electroweak theory, tackled the problem.

In that the electroweak interaction involves the leptons, and the strong interaction involves the quarks, a unified theory must imply that leptons and quarks have a basic, underlying similarity; that, under some conditions, one can be turned into the other. The natural assumption is that quarks can be turned into leptons, because this would be the direction of declining mass and energy.

Suppose, then, that a quark inside a proton is converted into a lepton. The proton would then no longer be a proton; it would have broken down into such less massive particles as kaons, pions, muons, and positrons (all positively charged, preserving the conservation of electric charge). The kaons, pions, and muons would eventually decay into positrons, meaning that, overall, protons would change into positrons.

This violates the law of conservation of baryon number.

However, all of the conservation laws are merely deductions from observations. We have never observed any change that alters the baryon number in an isolated system, so we naturally *assume* that such a change can never take place—and that gives us the conservation law. Nevertheless, however powerful and convenient the conservation laws are, they remain assumptions, and scientists must be ready now and then to accept the fact that a given conservation law might not work under all conceivable circumstances. They found this to be so in the case of the law of conservation of parity, as I explained earlier.

Still, scientists have been studying protons intensively for many decades, and no proton has ever been seen to decay. On the other hand, because scientists are quite convinced it cannot decay, they haven't put an emphasis on finding out for sure.

In addition, the unification of interactions that are now extant (there are several varieties) indicate that the half-life of the proton is extremely long. It would take 10^{31} years (ten million trillion trillion years) for half the protons in any given sample of matter to break down. In that the Universe is only about 15 billion years old, the half-life of the proton would be nearly 70 billion trillion times the age of the Universe. The number of protons that have broken down in the course of the entire lifetime of the Universe would thus be an insignificant fraction of the whole.

But it would not be zero! If you start with 10^{31} protons, which is what you would find in a tank holding some 20 tons of water, there would be an even chance of having one proton break down in the course of a single year. Detecting that one proton in the 20 tons of water and identifying its breakdown as due to the change of a quark into a lepton would not be an easy job, and scientists, who made some initial attempts to investigate the subject, have not yet succeeded in spotting such a breakdown.

Success or failure is important. Success will go a long

way toward establishing the validity of the interaction unification, the so-called grand unified theory; failure would
cast it into doubt.

Then, too, think of the light it would cast on the fate
of the Universe. If the Universe is open and expands forever, it will very slowly lose its proton content. It will
eventually become an unimaginably vast and thin cloud of
leptons—electrons and positrons (and, of course, photons
and neutrinos).

Of course, we also suspect that as the Universe ages
more and more of it will be concentrated into black holes—
and we haven't the faintest idea what the laws of nature
are like at the center of black holes. Will there be hadrons
of some sort at these centers? Will they decay, very, very
slowly, but very, very surely, and will the black holes eventually disappear? The puzzles continue—and will probably
continue forever.

The Beginning of the Universe

The Universe is at present expanding. Regardless of
whether it is open or closed, it is *at present* expanding. This
means it was smaller last year than it is now, and smaller
still the year before, and so on.

If we look into the future, it is at least conceivable that
there is unendingness about it, for the Universe might be
open, and might expand forever. If we look back at the
past, however, there is no chance of unendingness. The
Universe grows smaller and smaller, and at some moment
in the far past it can be viewed as having shrunk down to
some minimum size.

The first person to point this out in some detail was
the Belgian astronomer Abbé Georges Henri Lemaitre

(1894–1966). In 1927, he suggested that, in looking backward, there was a time when the matter and energy of the Universe were literally squashed together into one exceedingly dense mass. He called it the cosmic egg, thinking of it as unstable. It exploded in what we can only imagine to have been the most gigantic and catastrophic explosion the Universe is capable of affording. The effects of that explosion are still with us in the form of the expanding Universe. The Russian-American physicist George Gamow (1904–1968) called this the big bang, and the name stuck.

Naturally, there was some resistance to the notion of the big bang. Other scenarios were advanced that would account for the expanding Universe. The issue was not settled until 1964, when the German-American physicist Arno Allan Penzias (b. 1933) and the American physicist Robert Woodrow Wilson (b. 1936) studied radio wave radiation emanating from the sky.

In no matter which direction they looked, if they penetrated far enough they would detect radiation that had been traveling for so many billions of years that it must have originated in the big bang itself, if there was one. They found a faint radio wave background of identical intensity from every part of the sky, which was taken to represent the distant "echo" of the big bang. Physicists accepted this as establishing the big-bang theory, and Penzias and Wilson received a Nobel prize in 1978 for their work.

The big-bang theory has its problems, of course. For instance, when did it happen? One way to determine this is to measure the rate at which the Universe is now expanding and then work backward, allowing for the intensification of the gravitational pull as the Universe becomes smaller and denser.

It is a great deal easier to say this than to do it. There are several ways in which the rate of expansion can be

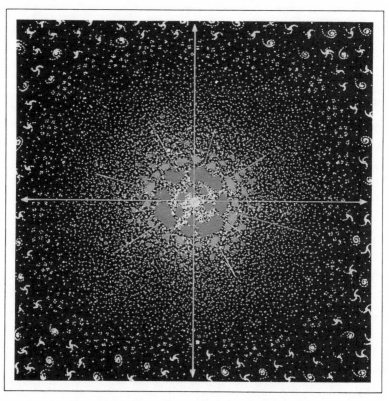

Today, radio astronomers can still listen to the distant "echo" of the big bang.

determined, in which the age of the oldest stars can be measured, and in which the distance of the farthest objects we can see can be determined (and, therefore, the amount of time it took radiation to reach us from these objects).

The results tend to conflict, and estimates of the time of the big bang vary from 10 to 20 billion years ago. Usually, people split the difference and assume that the Universe is, counting from the moment of the big bang, 15 billion years old, but I suspect that the truth is closer to 20 billion.

There are other more subtle difficulties. The radio wave background that Penzias and Wilson detected is extremely uniform in all parts of the sky, and represents an overall (average) temperature of the Universe of three degrees above absolute zero. This is puzzling, because to have a situation in which the temperature is the same everywhere, there usually has to be contact of some sort between the various parts so that heat can flow from here to there, evening out. This can't have happened in the Universe because different parts of it are separated by a greater distance than light can travel in the course of the entire lifetime of the Universe. Nothing else can travel faster than light, so what has evened out the temperature? What, in other words, makes the Universe so smooth?

Another problem is just the reverse. If the Universe *is* smooth, why didn't it stay smooth? Why isn't it just a featureless blob of subatomic particles, expanding endlessly? Why did the particles condense into huge pieces that became clusters of galaxies, with the galaxies condensing into individual stars. In other words, why is the Universe so smooth in some ways and so lumpy in others?

There are some other problems, too, but all of them— the age, the smoothness, the lumpiness, and so on—depend on just what happened at the very beginning of the Universe in the first instants after the great explosion. Naturally, no one was there to see, but scientists try to reason it out from what they know of the present state of the Universe, and from what they have learned concerning subatomic particles.

Thus, they assume that as they move backward in time, closer and closer to the big bang, the temperature grows higher and higher and the energy density is more and more enormous. Scientists feel that they cannot talk of times less than 10^{-45} second (one billion of a trillionth of a trillionth of a trillionth of a second) after the big bang. Less time than that brings about conditions so extreme that space and time themselves have no meaning.

However, the Universe cools rapidly in incredibly short fractions of a second. At first it was nothing but a sea of quarks, which existed freely because nothing else could exist, and because they had too much energy to even settle down enough to combine with each other.

By the time the Universe was a millionth of a second old, however, the quarks had split up into present-day quarks and leptons, and those quarks had cooled sufficiently to be able to combine so that the baryons and mesons were formed. Free quarks were never seen again. The interactions, which started as only one form, were splitting apart into the four we now recognize. When the Universe was one second old, it had rarefied to a point at which neutrinos had stopped interactions with other particles. They had begun to exist in free indifference of the rest of the Universe, and have done so ever since. Once the Universe was three minutes old, the simpler atomic nuclei began to be formed.

After a hundred thousand years, electrons began to circle nuclei. Atoms were formed. Thereafter, matter began to condense into galaxies and stars, and the Universe began to take on the shape we know.

Still, scientists could not prevent themselves from thinking of time zero, the actual instant of the big bang before the 10^{-45} limit. Where did the material of the cosmic egg come from?

If we consider the situation as it was before the cosmic egg was formed, we might visualize a vast illimitable sea of nothingness. Apparently, though, that is not an accurate description of what would exist. The nothingness contains energy. It is not quite a vacuum because, by definition, a vacuum contains nothing at all. The pre-Universe, however, had energy, and although all of its properties were otherwise those of a vacuum, it is called a false vacuum. Out of this false vacuum, a tiny point of matter appears where the energy, by the blind forces of random changes, just happens to have concentrated itself sufficiently for the purpose. In fact, we might imagine the illimitable false vacuum to be a frothing, bubbling mass, producing bits of matter here and there as the ocean waves produce foam.

Some of these bits of matter might disappear promptly, subsiding into the false vacuum from which they came. Some, on the other hand, might be large enough, or have been formed under such conditions generally, as to undergo a rapid expansion in a way that makes certain the Universe will form and survive, possibly, for many billions of years.

It might be, then, that we inhabit one of an infinite number of Universes in various stages of development and, for all we know, with different sets of laws of nature. However, there is absolutely no way of communicating with any other Universe, and we are forever confined to our own as a quark is confined to a hadron. This should not plague us unnecessarily. Our Universe is large enough all by itself, and varied enough, and puzzling enough for all purposes.

Viewing such a beginning of the Universe, the American physicist Alan Guth suggested in 1980 that in the very early stage of the Universe there was a rapid "inflationary" phase. This picture is described as the inflationary Universe.

It is difficult to grasp how brief the inflationary period is and how enormous is the inflation. The inflation starts about 10^{-35} second (ten trillionths of a trillionth of a trillionth of a second), with the Universe doubling in size for every 10^{-35} second thereafter. After a thousand doublings (only 10^{-32} second after the big bang), the inflation ceases. This difference in time (ten billionths of a trillionth of a trillionth of a second) was enough, however, to enable the Universe to grow 10^{50} times in volume. It ends the inflationary period with a hundred trillion trillion trillion trillion times the volume it had at the start. Moreover, in increasing its volume it incorporated more of the false vacuum and its energy content, thereby increasing its mass enormously. It can be shown that such a rapid initial inflation was what made the Universe smooth, and just about flat, possessing just the mass density that would place it on the boundary between being open and closed.

Guth's inflationary Universe didn't explain all properties of the present-day Universe. Scientists have been working to modify it so that it can give a better picture of what now surrounds us, especially as concerns the formation of the galaxies.

In order to do this, there must be a further unification. Not only must the strong and electroweak interactions be brought under a single umbrella, but gravitational interaction as well. Gravitation has so far resisted all attempts to incorporate it, but scientists are working with something called superstring theory, which they also call the "theory of everything."

It is not only that baryons and leptons are brought together as two different examples of something more fundamental, but fermions and bosons are unified and considered two different examples of something more fundamental. A new group of particles has been postulated in which there are new fermions analogous to our bosons, and new bosons analogous to our fermions.

Where this will go, I cannot say. There seems no point in trying to outline the current thinking, for it is bound to change and to be modified almost from day to day. Moreover, none of it has any observational backing at all, so that it remains merely speculation.

Still, there is the dream of a single set of equations that can cover all of the particles that exist in the Universe, as well as all of the interactions that involve them. This we need for a firm picture of a Universe that began with a single type of particle governed by a single type of interaction—a particle that as it gradually cooled, divided itself into the grand variety of effects we experience today.

And it all began with some ancients who questioned how far one could divide matter. It shows what asking the right questions can bring about.

INDEX